A Second Course of Electricity

These fundamental things have got to be simple
Lord Rutherford

A.E.E.McKENZIE,M.A.

A second course of electricity

Third edition revised by

J.JENKINS,B.Sc., M.Inst.P.
Lecturer in Electrical Engineering
South Devon Technical College
and
W.H.JARVIS,M.A.,M.Inst.P.
Head of Physics, Rannoch School

CAMBRIDGE
AT THE UNIVERSITY PRESS
1973

Published by the Syndics of the Cambridge University Press
Bentley House, 200 Euston Road, London NW1 2DB
American Branch: 32 East 57th Street, New York N.Y.10022

© Cambridge University Press 1973

ISBN: 0 521 08628 0

Printed in Great Britain
by William Clowes & Sons, Limited
London, Beccles and Colchester

CONTENTS

PORTRAITS

PUBLISHERS' NOTE

This book is a thorough revision of the late Mr A. E. E. McKenzie's *A Second (M.K.S.) Course of Electricity*, first published in 1956. In his preface to the first edition Mr McKenzie expressed the hope that the book would help schools to appreciate the merits of the m.k.s. system, and that the use of m.k.s. units would offer many advantages. The aims expressed then have been largely achieved, and the currently accepted SI system of units is a development of the m.k.s. system which retains the advantages which Mr McKenzie saw with m.k.s.

The Publishers would also like to repeat their thanks to those people who helped Mr McKenzie with the earlier editions of the book, notably Mr H. Tunley and Mr H. Gervis. Their contributions continue to be of value in this new edition.

PREFACE
TO THE THIRD EDITION

The 1966 Report on the Teaching of Electricity, published by John Murray for the Association for Science Education, recommended a different approach to the teaching of quantitative electrostatics and magnetism, the main point being that the vectors E and B alone are used, D and H being dropped. This edition conforms mainly to the 1966 report.

Units and abbreviations have been revised in great detail, to conform to SI and British Standards recommendations. The revisers feel, however, that the gram(me) and the centimetre will always be with us, so these have not been altered.

There has been and always will be some controversy over the choice of negative indices or the solidus. The revisers feel that anyone taking Physics at Advanced level must be familiar with the solidus, the word 'per', and negative indices, as he will meet all three in everyday life. Accordingly, all three are used freely in the text.

The revisers are very grateful for the detailed and careful help they have had from W. F. Archenhold, Department of Education, University of Leeds, and from the son of the author, Nigel McKenzie, Research Institute for Fundamental Physics, Kyoto University, Japan.

J.J.
W.H.J.

ACKNOWLEDGEMENTS

Thanks to the following for permission to use photographs and diagrams: Dr R. M. Bozorth and Monsieur J. Langevin (Fig. 5.15, from the *Journal de Physique*, **12**, 308, 1951); Dr R. J. Holt (Fig. 13.33); Professor C. F. Powell (Fig. 13.35); The Controller, H.M. Stationery Office and the Hydrographer of the Navy (Figs. 6.8 and 6.11); SIFAM (Fig. 2.4); Exide Batteries Ltd (Fig. 7.1); Mallory Batteries Ltd (Figs. 7.2 and 7.3); Alkaline Batteries Ltd (Fig. 7.4); Chloride Technical Centre, Manchester (Fig. 7.5).

The portraits of Ampère, Galvani and Volta are from the Library of the Royal Society. Those of Maxwell and Lord Rutherford are from portraits in the National Portrait Gallery. The portrait of Sir J. J. Thomson is taken from W. Nicholson's painting, at Trinity College, Cambridge. That of Coulomb is from the portrait by Emile Lecomte, in the Musée de Versailles. That of Faraday is from a lithograph of 1851 by S. H. Maguire. That of Lord Kelvin is from a photograph dated 1906, at the University Press, Cambridge. The portraits of Weber, Henry, Hertz, Joule and Ohm are by courtesy of Picture Post Library. Those of Einstein and Niels Bohr are by courtesy of Camera Press, Ltd.

We are very grateful to the Examining Bodies for allowing us to reproduce questions. The following is the key to the acknowledgements made after each question: (OS) Oxford Scholarship; (CS) Cambridge Scholarship; (O & C) Oxford and Cambridge; (O) Oxford Local; (C) Cambridge Local; (L) London; (N) Northern Universities; (B) Bristol; (D) Durham General Certificate of Education and Higher School Certificate Examinations.

SI Units

The International System of Units (SI) is a coherent system comprising the six Basic Units. All others are derived from these.

Basic Units	Physical quantity	Usual symbol	SI unit	SI unit symbol
	Length	l	metre	m
	Mass	m	kilogramme	kg
	Time	t	second	s
	Electric current	I	ampere	A
	Thermodynamic temperature	T	kelvin	K
	Luminous intensity	I	candela	cd

Note: another unit, called the 'mole', is about to be adopted. See p. 149.

Named Units	Physical quantity	Usual symbol	SI unit	SI unit symbol	Derivation
	Force	F	newton	N	$kg\ m\ s^{-2}$
	Energy	W	joule	J	$N\ m$
	Power	P	watt	W	$J\ s^{-1}$
	Electric charge	Q	coulomb	C	$A\ s$
	Electric potential p.d., e.m.f.	V E	volt	V	$J\ C^{-1}, W\ A^{-1}$
	Capacitance	C	farad	F	$A\ s\ V^{-1}, C\ V^{-1}$
	Resistance	R	ohm	Ω	$V\ A^{-1}$
	Frequency	f	hertz	Hz	s^{-1}
	Inductance	L	henry	H	$V\ s\ A^{-1}$
	Magnetic flux	ϕ	weber	Wb	$V\ s$
	Magnetic flux density	B	tesla	T	$Wb\ m^{-2}$
	Pressure	P	pascal	Pa	$N\ m^{-2}$

	Physical quantity	Usual symbol	SI unit	SI unit symbols
Other Units	Area	A	square metre	m²
	Volume	V	cubic metre	m³
	Density	ρ	kilogramme per cubic metre	kg m⁻³
	Velocity (linear)	v	metre per second	m s⁻¹
	Acceleration (linear)	a	metre per second per second	m s⁻²
	Torque	T	newton metre	N m (see p. 2)
	Electric field intensity	E	volt per metre	V m⁻¹
	Electric flux density	D	coulomb per square metre	C m⁻²
	Electro-magnetic moment	M	ampere . metre squared	A m²

	Power of 10	Name	Symbol
Powers of Ten Notation	$10^9 = 1\ 000\ 000\ 000$	giga	G
	$10^6 = 1\ 000\ 000$	mega	M
	$10^3 = 1000$	kilo	k
	$10^{-3} = 0{\cdot}001$	milli	m
	$10^{-6} = 0{\cdot}000\ 001$	micro	μ
	$10^{-9} = 0{\cdot}000\ 000\ 001$	nano	n
	$10^{-12} = 0{\cdot}000\ 000\ 000\ 001$	pico	p

The student should already be familiar with the powers of ten notation and most of the mechanical units. As each new unit is introduced in the text the student should make sure it is memorized so that by the time he finishes the text he is fully familiar with the contents of these tables.

The following examples should help to make this clear:

(a) 1 632 000 000 volts = 1·632 GV
(b) 5 900 000 ohms = 5·9 MΩ
(c) 1971 metres = 1·971 km
(d) 0·027 coulomb = 27 mC
(e) 0·000 012 amperes = 12 μA
(f) 0·000 000 589 metre = 589 nm
(g) 0·000 000 000 093 second = 93 ps

1   The electric current

Mechanical units

Electrical units are derived from mechanical units, and hence we begin the quantitative study of electricity with the definition of the mechanical units concerned.

Three primary units are defined arbitrarily and the remainder of the mechanical units are deduced from them. The primary units in the metric system are the metre, the kilogramme and the second. The metre was formerly defined as the distance between two marks on a platinum iridium bar, kept at Sèvres in France, while the second was defined as a certain fraction of the mean solar day. Advances in experimental techniques have lead to the redefinition of the metre and the second in terms of atomic quantities which can be measured very accurately.

The *metre* (*m*) is now defined as being equal to 1 650 763·73 wavelengths in a vacuum of the orange line (spectroscopic designation $2p_{10} - 5d_5$) emitted by the krypton-85 atom.

The *kilogramme* (kg) is the unit of mass, and is the mass of a platinum–iridium cylinder kept at Sèvres.

The *second* (s) is the unit for time interval, and is the duration of 9 192 631 770 cycles of the radiation corresponding to the transition between the two hyperfine levels of the ground state of the caesium-133 atom.

Unit of force

Unit force produces unit acceleration when applied to unit mass. The SI unit of force is the *newton* (N).

1 newton gives to a mass of a kilogramme an acceleration of 1 m s^{-2}.

The size of a newton can be judged from the fact that 1 newton is about the weight of an apple!

The unit of force on the centimetre–gramme–second (c.g.s.) system of units, which will not be employed in this book but which was in common use, is the dyne.

1 dyne gives to a mass of 1 g an acceleration of 1 cm s^{-2}.

Since 1 kg = 10^3 g
and   1 m = 10^2 cm
1 N = 10^5 dynes.

Unit of work, energy, and quantity of heat

The *joule* (J) is the work done by a force of one newton when its point of application is moved through a distance of one metre in the direction of the force. So we can write,

1 joule = 1 newton-metre.

In the SI system the joule is used for the measurement of *every* kind of energy.

(Torque, or turning moment, is also measured in newton-metres. However, in this case the force and the distance by which it is multiplied are at right angles and the units should not be confused with joules. A torque of one newton-metre allowed to turn through 2π radians does 2π joules of work.)

The unit of work in the c.g.s. system is the erg, and 1 erg = 1 centimetre-dyne.

Since 1 newton = 10^5 dynes
and   1 metre = 10^2 cm
1 joule   = 10^7 ergs.

Unit of power

Power is defined as the *rate* of doing work. The SI unit of power is the *watt* (W) and 1 watt = 1 joule per second. (The horse-power is 746 watts.)

Definition of the ampere

Two like parallel currents tend to attract each other and two unlike parallel currents to repel (Fig. 1.1); this can be demonstrated by setting up two parallel wires and passing currents through them. The phenomenon is used to define the unit of current, called the *ampere*, as follows:

The ampere is the strength of that constant current which, flowing through two parallel, straight and very long conductors of negligible cross-section, and placed in vacuo at a distance of 1 metre from each other, produces between these conductors a force of 2×10^{-7} newton per metre of their length.

CHARLES AUGUSTIN DE COULOMB   ANDRÉ MARIE AMPÈRE

CHARLES AUGUSTIN DE COULOMB (1736–1806) *was the first to apply
mathematics to the phenomena of electricity and magnetism. His early experiments on
the twisting of fine wires and invention of the torsion balance were put to use in his
investigations resulting in the discovery of the inverse square law in electrostatics. His
experimental skill was of the highest order; the forces he had to measure were small
and apt to be evanescent because of faulty insulation. Coulomb was an officer in the
French army and spent some time supervising the building of the fortifications in the
island of Martinique. During the Revolution he retired to his estate, but was later
recalled to Paris by Napoleon.*

ANDRÉ MARIE AMPÈRE (1775–1836) *was called by Maxwell 'the Newton
of electricity'. Ampère heard of Oersted's discovery, that an electric current in a wire
causes a magnetic compass to set at right angles to the wire, by letter in Paris on
11 September 1820 and in a short time he created a new subject, electromagnetism.
He formulated a clear idea of electric current – Oersted had spoken of the 'electric
conflict' in the wire – and also of electromotive force, coming quite near to a discovery
of Ohm's law. He discovered that a force exists between parallel wires carrying currents, that a solenoid behaves like a magnet and can be used to magnetize iron rods,
and put forward the suggestion that all magnetism might be due to electric currents.
Ampère was a man of peculiar temperament, religious, often tortured by doubts. As
a young man he suffered two severe blows, his father being guillotined, and his young
wife dying shortly after their marriage.*

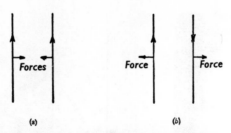

Fig. 1.1. (a) Like currents attract. (b) Unlike currents repel.

The symbol for ampere is A.

The value of the force, 2×10^{-7} newton, may seem arbitrary and surprising, but it results from the fact that the ampere was determined originally from the force exerted on a magnetic pole rather than on a wire carrying another current.

An ampere is a rate of flow of electricity. The unit of quantity of electricity, or of electric charge, is the *coulomb*. *1 coloumb is that quantity which passes in 1 second through any cross-section of a conductor in which a current of 1 ampere is flowing:*

$$\text{coulombs} = \text{amperes} \times \text{seconds}.$$

The symbol for coulomb is C.

The absolute measurement of the ampere, i.e. its measurement in terms of its definition, is performed by means of a current balance, which measures the force between neighbouring coils carrying the current. The current balance is expensive and demands considerable care durings its use; it is therefore not often to be found in the normal laboratory. With its aid very accurate determinations have been made of the mass of silver (or copper) deposited in electrolysis by 1 A in 1 s. This enables ammeters to be checked simply and very accurately by means of an instrument for measuring the mass of silver or copper deposited in electrolysis, known as a voltameter.

Potential difference

For an electric current to pass through a wire a potential difference (p.d.) must be applied across its ends. The p.d. is measured in volts (V).

We can compare an electric current, i.e. the movement of an electric charge between two points at different potentials, to the fall of a body between two points at different heights above sea-level. When a body of weight w (in newtons) falls a vertical distance s (in metres), the work done, or potential energy transformed, is ws (in newton-metres or joules). If the body were falling *in vacuo* it would acquire a kinetic energy of *ws* (in joules), but if it were falling in a viscous liquid like treacle, finally coming to rest, all the potential energy would be transformed into heat.

Similarly, when a charge of *q* (in coulombs) falls through a p.d. of *v* (in volts), the work done, or electrical energy transformed, is *qv* (in joules): this follows, as we shall see, from the definition of the volt. If the charge, consisting of a number of electrons, were passing through a vacuum, the electrical energy would be transformed into the kinetic energy of the electrons. But if the charge were passing through a wire, the electrical energy would be transformed into heat; the kinetic energy of the electrons is transferred, by bombardment, to the atoms of the wire.

Definition of the volt. Two points are at a p.d. of 1 volt if 1 joule of electrical energy is transformed during the passage of 1 coulomb of electricity between the points:

$$\text{coulombs} \times \text{volts} = \text{joules.}$$

Thus p.d. is energy or work, per unit charge.

An alternative way of defining the volt is in terms of the watt:

Two points are at a p.d. of 1 volt if power is expended at the rate of 1 watt when a current of 1 ampere is flowing:

$$\text{amperes} \times \text{volts} = \text{watts.}$$

The conservation of energy

From the definition of the volt it follows that if a p.d. of one volt applied across a circuit causes a current of one ampere to flow then one joule of electrical energy is dissipated in the circuit each second. The law of Conservation of Energy tells us that the electrical energy dissipation must be exactly equal to the sum of alternative forms of energy generated by the circuit. The alternative forms might be, for example, mechanical energy produced by a motor, light energy produced by a fluorescent lamp, or simply heat generated in the circuit.

Ohm's law for metallic conductors

The distinction and relation between p.d. and current was discovered by Ohm, who was guided by the analogy of the flow of heat along a conductor due to a difference of temperature.

The current through a conductor is proportional to the p.d. between its

GEORGE SIMON OHM   JAMES PRESCOTT JOULE

GEORGE SIMON OHM (1789–1854), *discoverer of the famous law which bears his name, was a teacher of mathematics and physics in a school in Cologne, where, working completely alone, he pursued his experimental investigations, using the skill learnt from his father, a locksmith, to construct his own apparatus. In 1826 an account of his experiments was published and also, in 1827, a paper entitled 'The galvanic circuit investigated mathematically', from which the experimental basis was omitted. Unfortunately only the latter attracted attention and was denounced as fantasy. Ohm lived in poverty and obscurity until the importance of his work was recognized in England and he was awarded the Copley Medal of the Royal Society. He achieved his ambition, a university professorship, a few years before his death.*

JAMES PRESCOTT JOULE (1818–1889) *was the son of a brewer at Salford, near Manchester, and had the good fortune to be taught science by John Dalton. His father built him a laboratory and he devoted his whole life to his researches. He was one of the first to formulate clearly the principle of the conservation of energy. At a meeting of the British Association in 1847 his paper on this subject would have attracted little attention had not William Thomson (Lord Kelvin) initiated a discussion by his striking and enthusiastic comments. Joule spent about forty years making determinations of the mechanical equivalent of heat by a number of different methods. He made his own galvanometer and graduated it with a voltameter. Resistances were measured in terms of a copper resistance by means of the galvanometer using Ohm's law.*

ends, *provided that physical conditions, such as the temperature, remain constant.*

Thus

$$\frac{V}{I} = \text{constant},$$

where V = p.d., I = current.

V/I can be used to define numerically what we intuitively regard as the resistance R of a conductor and Ohm's law becomes

$$\frac{V}{I} = R.$$

The practical unit of resistance is the *ohm*; it is *the resistance of a conductor through which a current of 1 ampere passes when a p.d. of 1 volt is maintained across its ends.*

The symbol for ohm is Ω.

It should be noted that Ohm's law is empirical, i.e. based on experiment. It holds with remarkable accuracy for metallic conductors under limited physical conditions.

The thermal effect of an electric current

When an electric current passes through a wire, the wire is heated. It was shown experimentally by Joule that the heat is proportional to I^2Rt, where I is the current, R is the resistance, and t is the time. This can be deduced, as follows, from the definition of the volt given on p. 5. Let V be the p.d. across the ends of the wire.

Charge which has flowed = It coulombs,

Electrical energy transformed = VIt joules   (definition of V)

and this must be identical with the heat energy produced so the heat generated

$$= VIt \text{ joules}$$
$$= I^2Rt$$
$$= \frac{V^2t}{R}.$$

Resistances in series

Two resistances R_1 and R_2, connected as in Fig. 1.2 (*a*), are said to be in series. Their combined resistance is equal to their sum:

$$R = R_1 + R_2.$$

It is clear that the same current flows through both resistances.

8

A SECOND COURSE OF ELECTRICITY

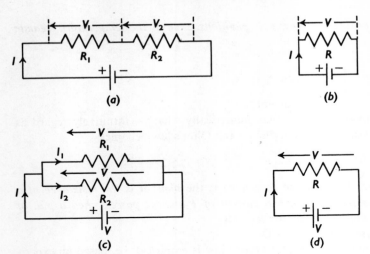

Fig. 1.2

The convention used throughout is that the p.d. between two points is marked by an arrow, the head representing the positive end.

The proof, using the Conservation of Energy, is as follows.

If the circuit in Fig. 1.2 (a) is to be electrically equivalent to the circuit in Fig. 1.2 (b), and by that we mean that R_1 and R_2 in series take the *same* current I from the *same* battery V as does a single resistor R, in a time t then

$$\text{heat developed in } R_1 \text{ and } R_2 = \text{heat developed in } R$$
$$V_1It + V_2It = VIt$$

(from which we note that $V_1 + V_2 = V$).

Alternatively, using Ohm's law in the form $V = IR$,

$$I^2R_1t + I^2R_2t = I^2Rt$$

or

$$R_1 + R_2 = R.$$

Resistances in parallel

When the resistances are connected as in Fig. 1.2 (c) they are said to be in parallel. Their combined resistance R is given by the formula

$$\frac{1}{R} = \frac{1}{R_1} + \frac{1}{R_2}.$$

In this case the current divides, the larger part going through the smaller resistance and the smaller part through the larger resistance.

It is clear in this case that the p.d. is the same across each component.

The proof of the formula, again using the Conservation of Energy, is as follows.

If the circuits shown in Figs. 1.2 (c) and 1.2 (d) are to be equivalent, then

$$\text{Heat developed in } R_1 \text{ and } R_2 = \text{Heat developed in } R$$
$$VI_1t + VI_2t = VIt$$

(from which we note that $I_1 + I_2 = I$)

or
$$\frac{V^2t}{R_1} + \frac{V^2t}{R_2} = \frac{V^2t}{R}$$

so
$$\frac{1}{R_1} + \frac{1}{R_2} = \frac{1}{R}.$$

If the conductance of a wire be regarded as the reciprocal of its resistance, $1/R$, the significance of the equation $1/R = 1/R_1 + 1/R_2$ becomes apparent. The total conductance of two wires in parallel is equal to the sum of their conductances.

Ohm's law applied to a complete circuit

A cell is said to produce an electromotive force (e.m.f.), which is measured by the total p.d. that it can produce, i.e. the p.d. that is maintained between its terminals when on open circuit.

When delivering a current, the cell must overcome not only the resistance of the external circuit, but also its own internal resistance. Thus Ohm's law, applied to a complete circuit, is

$$\frac{E}{I} = R + r,$$

where E volts = e.m.f. of the cell, I amperes = current, R ohms = external resistance, r ohms = internal resistance of the cell, or source.

This equation may be written as follows:

$$EIt \qquad\qquad = I^2Rt \qquad\qquad + I^2rt$$

or

$$\text{Electrical energy supplied by source} = \begin{array}{l}\text{Heat produced in external resistance}\end{array} + \begin{array}{l}\text{Heat produced in internal resistance.}\end{array}$$

A graph of V against I produces a straight line of negative slope r and whose intercept on the V axis is E.

IR is the p.d. between the terminals of the cell; denoting this by V, $V = E - Ir$. Thus the p.d. between the terminals of a cell drops when it delivers a current and 'the lost volts' Ir, drive the current through the internal resistance of the cell.

EXAMPLE

A cell of e.m.f. 1·5 volts and internal resistance 1 ohm is connected to a wire of resistance 4 ohms. Find the current flowing and the drop in p.d. between the terminals of the cell.

Applying Ohm's law to the complete circuit

$$I = \frac{E}{R+r} = \frac{1·5}{4+1} = 0·3 \text{ ampere.}$$

Applying Ohm's law to the 4 ohm resistance

$$V = IR = 0·3 \times 4 = 1·2 \text{ volts.}$$

Thus the p.d. across the 4 ohm resistance is 1·2 volts; this is also the p.d. between the terminals of the cell, since the ends of the 4 ohm resistance are connected to them. The drop in p.d. between the terminals of the cell when it delivers a current is therefore $1·5 - 1·2 = 0·3$ volt. Alternatively,

$$\begin{aligned} \text{Lost volts} &= Ir \\ &= 0·3 \times 1 = 0·3 \text{ volt.} \end{aligned}$$

The simplest way of finding the internal resistance of a cell is to measure the drop in p.d. with a voltmeter when a known current, as measured by an ammeter, is taken from the cell (see Fig. 1.3).

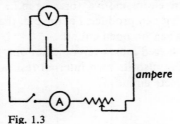

Fig. 1.3

The internal resistance is obtained from the equation

$$r = \frac{\text{drop in p.d.}}{\text{current causing that drop}}.$$

Arrangement of cells

A number of cells is called a battery. Cells may be connected in series or in parallel.

If n cells are connected in series (Fig. 1.4 (a))

$$\frac{nE}{I} = R + nr.$$

If m cells are connected in parallel (Fig. 1.4 (b))

$$\frac{E}{I} = R + \frac{r}{m}.$$

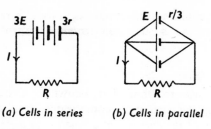

(a) Cells in series   (b) Cells in parallel

Fig. 1.4

Cells are seldom connected in parallel, but it is of interest to note that a battery gives its maximum power when its constituent cells are arranged so that its internal resistance is equal to the external resistance. This is not the most economical arrangement, however, since just as much energy is wasted inside the battery as is liberated in the external circuit.

The foregoing is, in fact, just one example of a general law, called the 'maximum power transfer theorem'. Simply stated it says that the electrical power transferred from a source to a load is always a maximum when the resistance of the load equals the internal resistance of the source.

Experimental verification of Ohm's law

To verify Ohm's law, the current I, through a fixed resistance R (see Fig. 1.5), is varied and measured by a current balance or ammeter, while the corresponding p.d.'s, V, across R are measured with an electrostatic voltmeter (p. 191). Then, if Ohm's law is true, $V/I =$ constant.

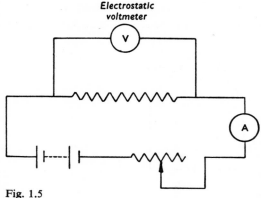

Electrostatic
voltmeter

Fig. 1.5

A current voltmeter could not be employed, since it is really a high-resistance galvanometer and is calibrated *assuming* the truth of Ohm's law!

Kirchhoff's laws

The following corollaries to Ohm's law, known as *Kirchhoff's laws*, are useful in calculating the currents in complicated circuits.

(1) *The algebraic sum of the currents at a junction in a circuit is zero,* i.e. $\sum I = 0$.

This means that there is no accumulation of charge at any point in the circuit.

(2) *In any closed circuit, the algebraic sum of the e.m.f.'s is equal to the algebraic sum of the products of the current and resistance in each part of the circuit, i.e.* $\sum E = \sum IR$.

This is an application of Ohm's law.

EXAMPLE

Calculate the currents in Fig. 1.6. The cell has a negligible internal resistance.

Kirchhoff's first law has been applied in the figure by inserting the currents, I_1, I_2, I_3, etc.

Applying Kirchhoff's second law to

$$ABDA, \quad 2I_1 + 1I_3 - 1I_2 = 0$$
$$BCDB, \quad 3(I_1 - I_3) - 2(I_2 + I_3) - 1I_3 = 0$$
$$FABCF, \quad 2I_1 + 3(I_1 - I_3) = 2.$$

The law can be applied also to *FADCF* but only three equations are required; moreover, this fourth equation can be derived from the above so that it contributes nothing new.

Solving the equations,

$$I_1 = \tfrac{16}{43}\,\text{A}, \qquad I_2 = \tfrac{30}{43}\,\text{A}, \qquad I_3 = -\tfrac{2}{43}\,\text{A}.$$

Since I_3 is negative it is clear that the arrow was marked in Fig. 1.6 in the wrong

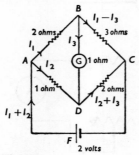

Fig. 1.6

direction; such an error is always automatically corrected by a minus sign in the solution.

What factors determine the steady temperature of a wire through which an electric current is flowing?

Assuming that for small excesses of temperature, the heat lost per second from the surface of a wire is proportional to θ, the temperature difference between the wire and its surroundings, show that $\theta \propto I^2\rho/r^3$, where I is the current, ρ is the resistivity of the material of the wire, and r is the radius of the wire.

The temperature of the wire depends upon the rate at which heat is generated in the wire and also upon the rate at which heat is lost from the surface. When a certain temperature is reached the two rates are equal and no futher rise in temperature occurs. The rate of loss of heat depends upon the area and nature of the surface of the wire, and, if Newton's law of cooling holds, upon the difference in temperature between the wire and its surroundings.

Let H = emissivity of the wire, i.e. the heat emitted per second per unit area per degree difference in temperature between the wire and its surroundings.

Resistance of wire per metre length = $\rho/\pi r^2$ (see p. 35).

Surface area of wire per metre length = $2\pi r$.

Heat generated in 1 metre of wire per second = Heat lost per second, i.e.

$$\frac{I^2\rho}{\pi r^2} = H2\pi r\theta.$$

$$\therefore \theta = \frac{1}{2\pi^2 H}\frac{I^2\rho}{r^3},$$

$$\theta \propto \frac{I^2\rho}{r^3}.$$

Electric power

Electric lamps and heaters are rated in watts to denote the rate at which they transform electrical energy. Since watts = amperes × volts, a 100 watt lamp on a 200 volt supply takes a current of $\frac{1}{2}$ ampere.

Electrical energy is sold by the kilowatt-hour, which is the energy transformed in 1 hour at the rate of 1000 watts, i.e. 1 kilowatt. The kilowatt-hour is sometimes called the Board of Trade Unit (B.T.U.) or just 'unit'.

Since 1 joule is 1 watt-second and 1 hour is 3600 s, 1 kilowatt-hour = 1000 × 3600 = 3 600 000 joules, or 3·6 MJ.

The transmission of electric power

It is preferable to transmit electric power along cables in the form of a small current at a high potential rather than a large current at a low potential, since the heat loss in the cable is smaller in the former case.

The maximum voltage at present used by the Grid is 400 kV.

The power loss in the cables is I^2R (in watts), where I is the current and R the resistance. Clearly this is small when I is small.

EXAMPLE

A d.c. dynamo, of output 20 kilowatts, supplies power through two cables each of resistance 1 ohm. Calculate the p.d. at the receiving end of the cables, and also the power loss in the cables, if the p.d. at the generating end is (*a*) 500 volts, (*b*) 10 000 volts.

$$(a) \text{ When generated at 500 volts, current} = \frac{20\ 000}{500}$$
$$= 40 \text{ A.}$$
$$\therefore \text{ Potential drop in the cables} = IR$$
$$= 40 \times 2 = 80 \text{ V.}$$
$$\therefore \text{ Potential difference at receiving end} = 500 - 80$$
$$= 420 \text{ V.}$$
$$\text{Loss in cables} = I^2R$$
$$= 40^2 \times 2$$
$$= 3200 \text{ W.}$$
$$(b) \text{ When generated at 10 000 volts, current} = \frac{20\ 000}{10\ 000}$$
$$= 2 \text{ A.}$$
$$\therefore \text{ Potential drop in the cables} = IR$$
$$= 2 \times 2$$
$$= 4 \text{ V.}$$
$$\therefore \text{ Potential difference at receiving end} = 10\ 000 - 4$$
$$= 9996 \text{ V.}$$
$$\text{Loss in cables} = I^2R$$
$$= 2^2 \times 2$$
$$= 8 \text{ W.}$$

Summary

Unit of current. One **ampere** is the strength of that constant current which, flowing through two parallel, straight and very long conductors of negligible cross-section, and placed *in vacuo* at a distance of 1 metre from each other, produces between these conductors a force of 2×10^{-7} newton per metre of their length.

Unit of charge. One **coulomb** is that quantity which passes in 1 second through any cross-section of a conductor in which a current of 1 ampere is flowing.

Unit of potential difference. One **volt** is the p.d. between two points if 1 joule of electrical energy is transformed when a charge of 1 coulomb flows between the points:

coulombs × volts = joules.
amperes × volts = watts.

Unit of resistance. One **ohm** is the resistance of a conductor through which a current of 1 ampere passes when a potential difference of 1 volt is maintained across its ends.

Ohm's law for a complete circuit:

$$\frac{E}{I} = R+r.$$

Heating effect = VIt.

Questions

1. How many of the following quantities must be known in order to determine the rest: current in amperes, p.d. in volts, resistance in ohms, time in seconds, energy in joules, power in watts, charge in coulombs? Write down the relations between these quantities.

2. Define the volt.
Electrons in a cathode tube are accelerated from rest through a p.d. of 2000 V. Calculate their resulting velocity.
(Charge on electron = $1 \cdot 60 \times 10^{-19}$ C, mass of electron = $9 \cdot 1 \times 10^{-31}$ kg.)

3. State Ohm's law and discuss its experimental verification.

4. A small arc lamp requires a p.d. of 45 V and takes a current of 5 A. What resistance must be placed in series with it if it is connected to a 110 V supply?

5. The terminals of a cell, of e.m.f. $1 \cdot 1$ V, are connected to a resistance of 2 Ω and the p.d. across them falls to 0·9 V. Calculate the internal resistance of the cell.

6. What different resistances can be made from three separate resistors of 1, 2 and 3 Ω?

7. Why is it that, in general, a voltmeter placed across the terminals of a cell gives a lower reading when the cell is giving current than when it is not?
Two identical cells are connected first in parallel and then in series to a resistance of 20 Ω. If the readings of a voltmeter connected across the resistance are 1·28 V and 1·60 V in the two cases, find the internal resistance of the cell and its e.m.f. on open circuit. (O & C)

8. Two coils, one of 100 Ω and the other of 200 Ω resistance, are connected in series with a 4 V battery of negligible resistance. A voltmeter, of resistance 200 Ω, is connected in parallel with each of the coils in turn. What voltage will it show in each case? (L)

9. The resistance of a telegraph wire between two stations is 55 Ω, the receiving

instrument has a resistance of 60 Ω and needs a minimum operating current of $\frac{1}{20}$ A. Find the smallest number of cells, each of e.m.f. 2·0 V and internal resistance 0·2 Ω, which, when joined in series, will enable a signal to be sent from one station to the other.

10. How would you arrange twelve similar cells, each of internal resistance 1 Ω, so as to send a maximum current through a wire of 0·70 Ω resistance ?   (C S)

11. An accumulator with an e.m.f. of 2 V and an internal resistance of 0·5 Ω, is connected in series with an external resistance R. Determine how the power developed in the external circuit varies with R by calculating this quantity when R has a number of known values, plotting the results and drawing a smooth curve through the points. Deduce the maximum value of the power which can be developed in the external circuit. Some suitable values of R are 0·1, 0·2, 0·4 and 0·8 Ω, but others may be needed.   (B)

12. A wire of resistance R is connected in series with a battery of e.m.f. V whose internal resistance is r. Find the rate of energy dissipation in the resistance R, and the condition for this to be a maximum.

Find the wattage and voltage of the brightest lamp which can be used with a $4\frac{1}{2}$ V battery of resistance 2 Ω.   (O & C)

13. State the laws of the heating effect of an electric current. Describe in detail an experiment to show how the rate of production of heat in a conductor during the passage of an electric current depends upon the strength of the current.

A 240 V electric kettle which has a heating element of resistance 80 Ω takes 14 minutes to heat $1\frac{1}{2}$ litres of water from 20 °C to boiling-point. What fraction of the heat generated is used to heat the water?   (N)

14. A recently marketed steam generator uses bare electrodes immersed in water and is served by the a.c. mains. Why would a d.c. supply be unsuitable? Calculate the mass of water at 100 °C vaporized per kilowatt-hour assuming that the specific latent heat of steam at this temperature is 2·26 MJ kg^{-1}.   (N)

15. How is electrical energy measured?

An undergraduate left his electric fire (240 V, 5 A) switched on all night (9 hours). The college fined him 25p. With electricity at 0·4p per unit did this cover the loss to the college?   (O S)

16 Making plausible assumptions about the quantities involved, estimate how much electrical energy is required to heat your morning bath.

Suggest a suitable power for an electric heater for this purpose and calculate for how long it would have to be switched on.   (C S)

17. In their determination of the specific heat capacity Callendar and Barnes used the apparatus in Fig. 1.7. A steady stream of water was passed through a tube and heated by the wire W carrying a current. Heat losses were reduced by surrounding the tube with a vacuum but, in order to eliminate losses, two experiments were performed. After performing one experiment the flow of water and the current were altered in such a way that the temperature of the outflowing water was unchanged. The rate of heat loss in the two experiments was then the same.

Suppose that, in the first experiment, the current in the wire is 2 A and the p.d. between its ends is 10 V, the rise in temperature of the water, flowing at 38·4 g per

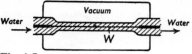

Fig. 1.7

minute, is 5·85 K. In the second experiment the p.d. is 12 V, the rate of flow 60 g per minute and the rise in temperature is unchanged. Calculate the specific heat capacity of water.

18. State and explain Kirchhoff's laws for the distribution of currents in a network of conductors.

Two batteries of e.m.f.'s 3 and 1 V respectively, and each of internal resistance 2 Ω, are connected in parallel and their terminals are joined by a resistance of 5 Ω. Determine the current in each branch of the circuit. (L)

19. The positive poles A and C of two cells are connected by a uniform wire of resistance 4 ohms, and their negative poles B and D by a uniform wire of resistance 6 Ω. The middle point BD is connected to earth. The e.m.f.'s of the cells AB and CD are 2 V and 1 V respectively, and their resistances 1 Ω and 2 Ω respectively. Find the potential at the middle point of AC. (O & C)

20. Two cells whose e.m.f.'s are 1·6 and 2·4 V have internal resistances of 2 and 4 Ω respectively. The two positive poles are joined by a wire of 6 Ω and the two negative poles by a wire of 8 Ω. If another wire of resistance 10 Ω is placed between the midpoints of these two wires, what is the p.d. between its ends?

(O & C)

21. State the laws which govern the distribution of current in a network of conductors.

Four wires, AB, BC, CD and DA, of resistance 2, 4, 2 and 4 Ω respectively, are connected in series. The positive pole of a cell of e.m.f. 2 V and internal resistance 1 Ω is connected to A and its negative pole to C. The positive pole of a cell of e.m.f. 1 V and internal resistance 2 Ω is connected to D and its negative pole to C. Find the current in each branch of the network, assuming that the resistances of the connecting wires may be neglected. (C S)

22. State and explain Kirchhoff's laws giving the division of current in a network.

A tetrahedron $ABCD$ is built up of wires each of resistance 2 Ω. The wires AB, BC, BD are of substance p and AD, AC, DC are of substance q, such that, if a circuit is made of p and q in series, and the junctions are kept at a temperature difference of 100 K, then an e.m.f. E is produced, tending to drive current from p to q across the hot junction. Calculate the currents in the wires of the tetrahedron if the points A and C are placed in ice, and B and D in boiling water. (C S)

23. A current of 10 A flows through a copper wire 0·166 cm diameter in a room whose temperature is 288 K. Calculate the steady temperature which the wire will finally attain. Assume that the emissivity of the surface of the wire is 10·4 W m^{-2} K^{-1}, that the emission of heat follows Newton's Law of Cooling and that the resistivity of copper is 2×10^{-8} Ωm. (D adapted)

24. It is desired to construct a 5 A fuse from tin wire which has a melting-point of 503 K and resistivity 22×10^{-8} Ω m at that temperature. Estimate the diameter of the wire required if the emissivity of its surface is 8.8 W m^{-2} K^{-1} excess temperature above the surroundings, whose temperature is 293 K. Neglect the heat lost by conduction along the wire.   (N adapted)

25. Derive an expression for the rate of generation of heat in a wire of resistance R traversed by a current I.

Calculate the heat radiated per square metre per second from the surface of the tungsten filament of an electric lamp in terms of the current, specific resistance, and radius. Hence, compare the filament diameters of a 60 W, 100 V lamp and a 100 W, 200 V lamp, assuming that the filaments are run at the same temperature.   (O & C)

26. In the 'Grid' system of electrical distribution, electrical energy is transmitted at 132 kV. Why is it advantageous to transmit energy at this high voltage and by alternating current, although the usual supply is 240 V?

The potential in a thundercloud is estimated to be about a million kilovolts, but the quantity of electricity discharged in one lightning flash is only about 20 coulombs. If the energy of the flash could be spread out over 24 hours, how many 5 h.p. machines could be kept working by it for this period? (1 h.p. = 746 watts.)   (C)

27. A battery of accumulators, each of e.m.f. 2.0 V and internal resistance 0.002 Ω, supplies energy to a factory through cables of total resistance 0.2 Ω. The supply voltage at the factory is 100 V and the power consumed there is 5 kW. Find the least number of cells which is necessary.   (N)

28. Discuss the advantages and disadvantages of the transmission of electrical power at high voltages.

A certain works consumes electrical energy at the average rate of 200 kW and the voltage of the supply (at the works) is 230 V. The resistance of the transmission line from the power station is 3×10^{-2} Ω. Calculate the loss in the cables. If the transmission were carried out at 2000 V in the same cables, being stepped up at the power station and down at the works by transformers, each having 99% efficiency, what would be the total energy loss in transmission?   (O & C)

29. Describe how you would investigate, over a fairly wide range of currents, the way in which the current through a metal wire maintained at constant temperature varies with the potential difference between its ends.

Using the simple electron-drift picture of conduction through a metal explain (a) why, under suitable condition, Ohm's law holds; (b) why the resistivity of a metal increases as the temperature rises; (c) why the passage of a current through a metal wire causes its temperature to rise.   (O 1968)

2 Electrical measurements

Methods of measuring current, potential difference, and resistance will be considered in this chapter.

Measurement of current

To determine the magnitude of a current, its magnetic effect, thermal effect, or chemical effect may be measured. The instruments used may be classified as follows:

(1) *Magnetic effect*, moving-magnet, moving-coil, moving-iron galvanometers.*

(2) *Thermal effect*, hot-wire galvanometer, thermocouple combined with moving-coil galvanometer.

(3) *Chemical effect*, voltameter.

The voltameter will be described in Chapter 7.

Moving-magnet galvanometer

A circular coil carrying a current produces a magnetic field which is perpendicular to its plane at its centre. A small magnet suspended at the centre of the coil tends to set itself in the direction of this field, perpendicular to the plane of the coil.

This is the principle of the galvanometer represented in Fig. 2.1. The magnet is suspended by a fine silk thread and carries a light pointer which moves over a scale. The galvanometer must be set before use so that the coil lies in the magnetic meridian; the magnet will then lie with its axis in the plane of the coil and the pointer will be in the middle of the scale. When a current is sent through the coil, its magnetic field will urge the magnet to turn E and W, but a controlling torque is provided by the earth's magnetic field which urges the magnet to remain N and S. The sensitivity of the galvanometer can be increased by reducing the value of the controlling magnetic field, by fixing a bar magnet at a suitable distance above the coil, with its N pole pointing north and with

---

* The term galvanometer is sometimes confined to instruments measuring small currents only; in this chapter it includes the movements of ammeters and voltmeters.

its axis lying north and south, so that it opposes the earth's field at the centre of the coil.

The sensitivity can also be increased, when designing the instrument, by reducing the radius of the coil and by increasing the number of turns. An increase in the number of turns, however, increases the resistance of the galvanometer, and, although the current sensitivity is increased, the voltage sensitivity, i.e. the deflection caused by unit p.d. across its terminals, is decreased. It is the voltage sensitivity that is concerned when the instrument is used for finding the balance point of a potentiometer or of a metre bridge.

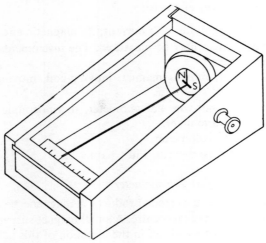

Fig. 2.1

This type of galvanometer has the merit of cheapness and it is suitable for detecting, but not for measuring, currents, being affected by stray magnetic fields in its neighbourhood.

Another moving-magnet galvanometer, known as a tangent galvanometer, which can be used for measuring currents but which is not now of practical importance, is described on p. 134.

Moving-coil galvanometer

A coil carrying a current behaves like a magnet, one face acting like a north pole and the other like a south pole (Fig. 3.4, p. 49). If its plane lies originally parallel to a magnetic field, it will tend to turn until its plane is perpendicular to the field, and it can be made to take up an

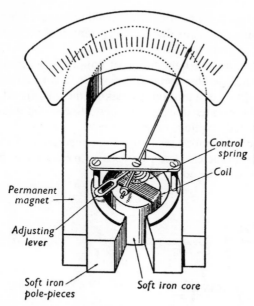

Fig. 2.2. Moving-coil galvanometer.

intermediate position by means of a mechanical controlling torque. This is the principle of the moving-coil galvanometer.

Figure 2.2 represents one form of the instrument. A coil of fine wire is pivoted on jewelled bearings between the poles of a permanent magnet and is attached to a light pointer moving over a circular scale. The coil swings round a fixed cylinder of soft iron, the purpose of which is to make the magnetic field converge towards the axis of revolution of the coil; the deflection is then directly proportional to the current. Two

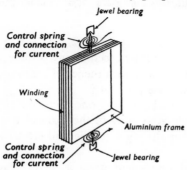

Fig. 2.3. Coil and control springs of moving-coil galvanometer.

control springs (Fig. 2.3) provide the controlling torque, which counter-balances the deflecting torque due to the current, and they serve also to lead the current in and out of the coil.

The coil is wound on a light frame of copper or aluminium. Eddy currents are induced in this metal frame while it is in motion causing 'damping' or resistance to motion, so making the instrument deadbeat; that is to say, the pointer does not oscillate like a pendulum but, on reaching its deflected position, remains at rest.

The moving-coil galvanometer is the most accurate and most widely used type of galvanometer. It can be used to measure a.c. by fitting a rectifier in series with the coil.

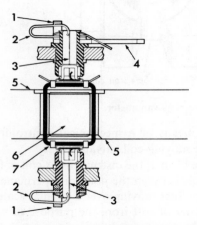

Fig. 2.4. Taut-band suspension. 1. Limiting stops; 2. tension springs; 3. gold alloy suspension strip (taut band); 4. zero adjustment arm; 5. magnet; 6. stationary core-piece; 7. aluminium damping former and coil.

A recent development in instrument technology is the use of a 'taut-band' suspension, which does away with the jewel bearings, and results in a very robust instrument. (See Fig. 2.4.)

The taut-band is in fact two ligaments, one each end of the moving coil. A special silver–copper–gold alloy in the form of a wire is rolled into a flat ligament, and has the right tensile and electrical resistance properties.

The use of a taut-band movement gives a meter several distinct advantages. It eliminates all the problems of wear, friction and susceptibility to shock that are associated with the conventional movement based on delicate jewelled pivots.

The mirror galvanometer

A very sensitive form of moving-coil galvanometer is represented in Fig. 2.5(a), known as a mirror galvanometer. The coil is suspended between the poles of a ring-shaped permanent magnet by means of a fine strip of phosphor-bronze, which provides a very much smaller controlling torque than the hair springs in Fig. 2.2. The current passes in and out of the coil through the supporting phosphor-bronze strip and through a similar strip below it, loosely coiled to eliminate torsion.

The deflection of the coil is measured by means of a beam of light reflected from a small concave mirror, M, and focused on a linear scale. In Fig. 2.5 (b), L represents a lamp directing a parallel beam of light

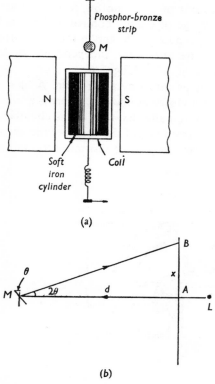

Fig. 2.5

perpendicular to the original position of the reflecting mirror, M. When M rotates through an angle θ, the reflected ray is turned through 2θ, and the spot of light on the scale moves from A to B.

$$\frac{x}{d} = \tan 2\theta = 2\theta, \quad \text{if } \theta \text{ is small and is measured in radians.}$$

Moving-iron galvanometer

There are two types of moving-iron galvanometer, the attraction type and the repulsion type.

The attraction type (Fig. 2.6) consists of a piece of soft iron mounted eccentrically on a spindle at the side of a coil through which the current to be measured is passed. The soft iron is attracted into the coil towards the region of stronger magnetic field just as a piece of iron is attracted towards a magnet. The controlling force in Fig. 2.6 is provided by gravity; when the pointer moves to the right the control weight moves to the left and the moment of the restoring force increases with the deflection. The disadvantage of gravity control is that the instrument cannot be tilted since this causes a change in the zero position of the pointer; the advantages are cheapness, independence of temperature and freedom from deterioration with age.

Damping in Fig. 2.6 is provided by air friction. A light aluminium piston moves in a chamber closed at one end. There is a small clearance (a few μm) between the piston and the walls of the chamber. When the piston moves towards the open end of the chamber the air pressure behind it is less than atmospheric, causing resistance to motion, until the

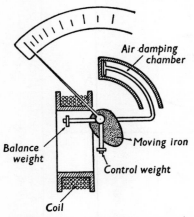

Fig. 2.6. Moving-iron galvanometer, attraction types.

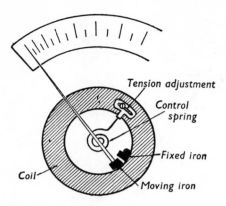

Fig. 2.7. Moving-iron galvanometer, repulsion types.

air has had time to pass through the clearance gap to equalize the pressures.

The repulsion type of moving-iron galvanometer (Fig. 2.7) consists of two rods of iron, one fixed rigidly in position and the other attached to a pivoted pointer, inside a coil carrying the current to be measured. Both rods become magnetized in the same direction and their mutual repulsion causes a deflection of the pointer. In Fig. 2.7 spring control is shown, but in some instruments gravity control is used.

Moving-iron instruments have two great advantages, they are cheap and they can be used to measure low frequency a.c. as well as d.c. The iron in Fig. 2.6 is attracted into the coil whatever the direction of the current; both rods in Fig. 2.7 are always magnetized in the same direction and they continue to repel each other even though the current is alternating. A marked disadvantage of the moving-iron instrument is the unevenness of its scale. It is liable to be affected by stray magnetic fields unless shielded by a soft-iron case. It is not as sensitive as the moving-coil galvanometer but its design has been greatly improved in recent years.

The hot-wire galvanometer

The hot-wire galvanometer (Fig. 2.8) consists of a fine wire, which is heated by the current to be measured, and a mechanism to enable the expansion or sag of the wire to cause movement of a pointer. The 'hot-wire' is made of platinum–iridium, which will withstand high temperatures without oxidation. To it is attached a phosphor-bronze

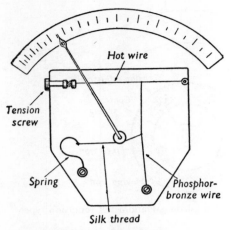

Fig. 2.8. Hot-wire galvanometer.

wire; this in turn is attached to a silk thread, kept taut by a spring, and passing round a pulley carrying the pointer.

The scale is uneven; the wandering of the zero, due to changes in room temperature, may be corrected by the tension adjustment. The power consumption is about four times that of the moving-coil galvanometer. The instrument is often used for radio-frequency alternating currents since it is independent of change in frequency or wave form.

Thermocouple instrument

Fig. 2.9 represents one type of thermocouple instrument. The current passes through a heater, the radiation from which warms the junction of a thermocouple connected in series with the coil of a moving-coil galvanometer. When the junction of the thermocouple is warmed a

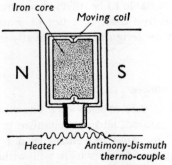

Fig. 2.9. Thermocouple instrument.

small current flows (see p. 221). The thermocouple may be in actual contact with the heater; the design of the instrument must then obviously be modified. The instrument is superior to the hot-wire galvanometer for radio-frequencies.

Determination of the sensitivity of a galvanometer

The most interesting and useful information about a galvanometer is its sensitivity, which may be expressed in divisions per microampere. To determine the sensitivity it is necessary to know the resistance of the galvanometer coil; this may be measured by the method given on p. 39.

The galvanometer is connected (Fig. 2.10) in series with a resistance box and an accumulator, whose e.m.f., E, has been determined by means of an accurate voltmeter. The deflection, θ, is measured with different, suitable resistances R in the box.

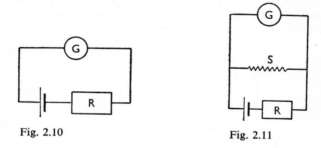

Fig. 2.10   Fig. 2.11

Then $I = E/(R+G)$, where G is the resistance of the galvanometer; I is plotted against θ, whence θ/I may be obtained.

If the galvanometer is very sensitive it may not be possible to obtain a sufficiently high resistance from the box. It is then necessary to shunt the galvanometer with a known resistance, S (perhaps $\frac{1}{10}$ ohm). Figure 2.11 shows the circuit; a reversing switch also may be included to enable deflections to be obtained in both directions in the galvanometer. The reader should verify that the current in the galvanometer is

$$\frac{SE}{R(G+S)+SG}.$$

Ammeters

The sensitivity of a galvanometer may be reduced by connecting in parallel with it a low resistance called a shunt. Part of the current then passes through the shunt instead of through the galvanometer. In this way a galvanometer can be converted into an ammeter. By means of a set of shunts of different resistances its range may be varied, for example, from 0–1·5 A, 0–5 A, 0–10 A, and so on.

A moving-coil galvanometer of resistance 5 Ω requires a current of 15 mA to produce a full-scale deflection. Calculate the resistance of the shunt necessary to convert it into an ammeter reading from 0 to 10 A.

In Fig. 2.12

p.d. across moving coil = p.d. across shunt, i.e.

$$0\cdot015 \times 5 = 9\cdot985\ S,$$

$$S = \frac{0\cdot075}{9\cdot985} = 0\cdot007\ 51\ \Omega.$$

The total resistance of the ammeter is less than that of the shunt; in this case it is less than 0·007 51 Ω. The resistance of an ammeter should always be low since otherwise, when placed in a circuit to measure the current, it might reduce the current considerably. Moving-iron, as well as moving-coil, galvanometers are often converted into ammeters.

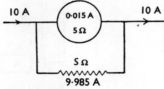

Fig. 2.12

Measurement of potential difference

The voltmeter

The obvious method of measuring the p.d. between two points is to connect them to a suitable galvanometer. Then, by Ohm's law, the p.d. is the product of the current in the galvanometer and the resistance of the galvanometer. The galvanometer must have a high resistance, since otherwise it will take a large current and lower appreciably the p.d. it is required to measure.

A voltmeter is a sensitive galvanometer connected in series with a high resistance.

A moving-coil galvanometer, of resistance 5 Ω, requires a current of 15 mA to produce a full-scale deflection. What resistance must be placed in series with it to convert it into a voltmeter reading 0–10 V?

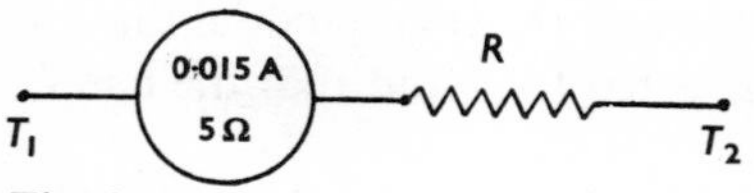

Fig. 2.13

Let the resistance in series with the galvanometer be R (Fig. 2.13).

When a p.d. of 10 V is applied across the terminals, T_1 and T_2, a current of 15 mA must flow to provide a full-scale deflection.

Applying Ohm's law

$$10 = 0.015 (5 + R)$$
$$\therefore R = 661.7 \ \Omega.$$

The series resistance of a voltmeter is usually fitted inside the case of the instrument; but sometimes a galvanometer is provided with a set of resistances which can be screwed to one of its terminals, thus enabling it to be converted into a voltmeter with different ranges. Both moving-coil and moving-iron voltmeters are in common use.

The potentiometer

Since the voltmeter just described must take some current, however small, it is bound to lower slightly the p.d. it is required to measure. There are two methods of measuring a p.d. without taking a current, the electrostatic voltmeter (see p. 191), and the potentiometer.

In its simplest form the potentiometer consists of a resistance wire of uniform cross-section through which a steady current is passed. In Fig. 2.14 AB represents the potentiometer wire and C the cell (usually an accumulator) supplying the steady current. There is a drop of potential down the wire from A to B; the p.d. between two points on the wire is proportional to their distance apart, and can be used to counterbalance an unknown p.d.

Thus, to compare the e.m.f.'s of two cells, one of the cells D_1 is connected as in Fig. 2.14 and the sliding contact is moved along AB until no current flows in the galvanometer. The p.d. between A and K_1 is

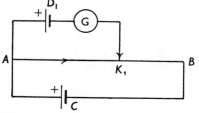

Fig. 2.14

then equal to the e.m.f. of the cell D_1. It is essential that the same poles of C and D_1 should be connected to A and that the e.m.f. of C should be greater than that of D_1.

The cell D_1 is now replaced by the second cell D_2 and the new position of the sliding contact, K_2, found.

$$\frac{E_1}{E_2} = \frac{AK_1}{AK_2}, \quad \begin{array}{l} E_1 = \text{e.m.f. of cell } D_1, \\ E_2 = \text{e.m.f. of cell } D_2. \end{array}$$

The Weston standard cadmium cell

A standard of p.d. must be employed with the potentiometer if the actual value of an unknown p.d. is to be determined. The standard used

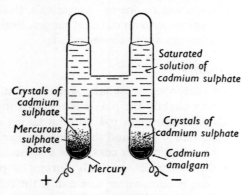

Fig. 2.15. Weston standard cadmium cell.

is the Weston standard cadmium cell shown in Fig. 2.15; when constructed to certain specifications it yields accurately 1·0183 volts at 293 K. No appreciable current should be taken from a standard cell or it will be ruined; it should therefore be connected in series with a high resistance when making preliminary adjustments with a potentiometer.

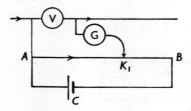

Fig. 2.16

Calibration of a voltmeter by a potentiometer

The potentiometer may be used to calibrate a voltmeter (Fig. 2.16). A suitable p.d. is put across the voltmeter by means of a cell and variable resistance – not shown in the figure. The sliding contact on the potentiometer wire is adjusted until no current flows in the galvanometer. Then the p.d. across the voltmeter is equal to the p.d. across AK_1. The voltmeter is now replaced by a standard cell of e.m.f., E, and the new position, K_2, of the sliding contact found.

$$\frac{\text{p.d. across voltmeter}}{E} = \frac{AK_1}{AK_2}.$$

The reading of the voltmeter is compared with this calculated value.

Calibration of an ammeter by a potentiometer

The potentiometer is very sensitive and easy to use. It is therefore employed in conjunction with a standard resistance and a standard cell, to calibrate ammeters.

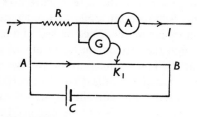

Fig. 2.17

In Fig. 2.17 R is the standard resistance, which is connected in series with the ammeter to be calibrated, so that a suitable current, I, can be passed through both. The sliding contact is adjusted until the galvanometer is undeflected; then the p.d. across R is equal to the p.d. across AK_1. The length, AK_2, of the potentiometer wire whose p.d. balances the e.m.f., E, of a standard cell is then found.

By Ohm's law, p.d. across $R = IR$.

$$\therefore \frac{IR}{E} = \frac{AK_1}{AK_2},$$

$$I = \frac{AK_1}{AK_2}\frac{E}{R}.$$

Knowing E and R and measuring AK_1 and AK_2 the current may be determined; its value is compared with the reading of the ammeter.

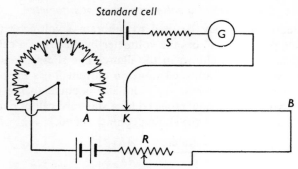

Fig. 2.18. Crompton potentiometer.

The Crompton potentiometer

A potentiometer, in order to be sensitive, must have a very long wire. This is obviated, however, in the Crompton potentiometer (see Fig. 2.18), by having in series with a short wire, AB (50 cm long), a number of coils each of resistance equal to that of the wire. Figure 2.18 shows the principle of the instrument. Suppose ten coils and a length equal to 0·18 of AB are made to balance a standard Weston cell of e.m.f. 1·018 V by varying the current through the instrument by means of R (actually incorporated in the instrument). This means that the voltage across each coil or AB is exactly 0·1 V and, if the wire is divided into 100 divisions, each division represents a millivolt. The current through the instrument is then left unaltered and unknown p.d.'s can be balanced against the requisite number of coils and part of AB. The resistance S is to protect the standard cell. This type of instrument is employed in many laboratories for calibrating voltmeters and ammeters.

Measurement of resistance

The Wheatstone bridge

The most used method of measuring resistance is the Wheatstone bridge. This has the great advantage of being a null method; that is to say adjustments are made until a galvanometer is undeflected and hence the result does not depend on the accuracy of an instrument. Reliable standard resistances are, however, required.

Four resistances P, Q, R, S are arranged as in Fig. 2.19. One of these, say P, is the unknown resistance; then Q must be a known standard resistance and the values of R and S or their ratio must also be known.

A sensitive galvanometer and a cell are connected as shown. If the

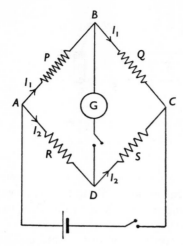

Fig. 2.19

resistances R and S are adjusted so that no current flows in the galvano-meter (when the bridge is said to be balanced), it can be proved that

$P/Q = R/S$,

whence P can be calculated.

Since no current flows through the galvanometer,

current through P = current through $Q = I_1$,

and current through R = current through $S = I_2$.

Also potential at B = potential at D.

∴ p.d. between A and B = p.d. between A and D,

i.e. $I_1 P = I_2 R$.

Similarly p.d. between B and C = p.d. between D and C,

i.e. $I_1 Q = I_2 S$,

dividing $P/Q = R/S$.

The metre bridge

The simplest experimental arrangement of the Wheatstone bridge is the metre bridge (Fig. 2.20). A wire AC of uniform cross-section and 1 metre long, made of some alloy such as constantan so that its resist-ance is of the order of 1 ohm, lies stretched between two thick brass or copper strips bearing terminals, above a metre ruler. There is another

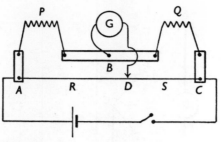

Fig. 2.20

thick brass strip bearing three terminals to facilitate connections and also a sliding contact D which can move along the metre wire.

Figure 2.20 has been lettered identically with Fig. 2.19 to show that the two circuits are similar.

The position of the sliding contact is adjusted until no current flows in the galvanometer. Then

$$\frac{P}{Q} = \frac{R}{S} = \frac{\text{length } AD}{\text{length } CD}.$$

$$\therefore P = Q\frac{AD}{CD}.$$

The Post Office box

Another practical form of the Wheatstone bridge is the Post Office box (Fig. 2.21), originally designed for locating faults in telephone cables.

The box consists of three sets of resistances which can be brought into circuit by removing plugs. P and Q are known as the ratio arms and each contains resistances of 10, 100 and 1000 Ω; R will give any resistance up to about 6000 Ω (in this particular box). Two switches are connected inside the box to the terminals indicated by the dotted lines. The circuit, connected as shown, is identical with that in Fig. 2.19, but in this case S is the unknown resistance to be measured.

When the circuit has been connected up the method of procedure is as follows:

Take out the 10 Ω plug from both P and Q. Suppose the galvanometer deflects to the left when $R = 1$ Ω and to the right when $R = 2$ Ω. Then the resistance of S lies between 1 and 2 Ω. Now make P 100 Ω and Q 10 Ω. R will lie between 10 and 20 Ω, say between 16 and 17. Finally, make P 1000 Ω and Q 10 Ω. R will lie between 160 and 170 Ω. Suppose 163 Ω produces a deflection of $\frac{1}{2}$ division to the left in the gal-

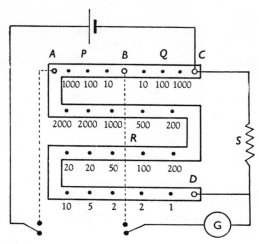

Fig. 2.21

vanometer and 164 Ω a deflection of $\frac{1}{2}$ division to the right. Then the resistance of S is 1·635 Ω. See that all the plugs in the box are tight while performing the experiment; otherwise unexpected resistances are introduced into the circuit.

Resistivity

To compare the resisting powers of different substances a term *resistivity* is used.

The resistivity of a substance is the resistance of a sample of the substance one metre in length and of cross-section one square metre.

If the length of a wire is doubled this is equivalent to two of the original lengths in series; hence the resistance is doubled. Thus the resistance R of a wire is proportional to its length, l. Again, if the area of cross-section of a wire is doubled, this is equivalent to two wires in parallel; hence the resistance is halved. Thus the resistance is inversely proportional to the area of cross-section A, i.e.

$$R \propto l/A,$$

If ρ = resistivity of the substance of which the wire is made,

$$R = \rho(l/A).$$

The resistivity of a substance is found by measuring the resistance of a known length of wire and also its diameter by means of a micrometer screw gauge. The units of resistivity are Ω m.

Electrical *conductivity* is the reciprocal of resistivity.

EXAMPLE

The resistance of 80·0 cm of constantan wire, whose diameter of cross-section is 0·457 mm, is 2·39 Ω. Find the resistivity of constantan.

$$\rho = \frac{RA}{l}.$$

$$l = 0·80 \text{ m}, \quad A = \pi \times \left(\frac{0·457}{2}\right)^2 \times 10^{-6} \text{ m}^2.$$

$$\therefore \rho = \frac{2·39 \times \pi \times 0·2285^2 \times 10^{-6}}{0·80}$$

$$= 4·90 \times 10^{-7} \text{ Ω m}.$$

The effect of temperature on resistance

The resistance of most substances increases when the temperature rises. The *temperature coefficient of resistance* is defined by the equation

$$R_t = R_0(1 + \alpha t),$$

where R_t = resistance at t °C, R_0 = resistance at 0 °C, α = temperature coefficient of resistance. The temperature coefficient of resistance of a substance in the form of a wire may be found by measuring its resistance at different temperatures. If resistance is plotted against temperature the graph will be found to be straight only over small ranges of temperature showing that the above equation is only approximately true. The value of α, which is $(R_t - R_0)/R_0 t$, may be obtained from the graph (see Fig. 2.22).

The temperature coefficients of resistance of most pure metals do not differ widely and are of the order of 0·004 per K which is roughly equal to $\frac{1}{273}$, the cubic expansivity of gases.

Experiments performed in 1911 and subsequently have shown that when rings of mercury, lead, tin and thallium are cooled to the temperature of liquid helium their resistances disappear almost entirely. A current once started in the ring will continue for days. The phenomenon is called *superconductivity*.

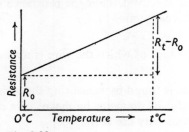

Fig. 2.22

When the temperature of a pure metal is raised from 273 K its resistance increases by approximately $\frac{1}{273}$ of its resistance at 273 K for each 1 K rise. Since the temperature of the filament of a gas-filled tungsten lamp when lit is over 3000 K, it follows that the resistance of the lamp when lit is more than ten times its resistance when cold.

The temperature coefficient of resistance of most alloys is considerably less than that of pure metals. The value for manganin is of the order of 0·000 05 and for eureka 0·000 01. These two alloys are used for making resistance coils, since a slight change of temperature has no appreciable effect on their resistance.

Carbon, boron, glass, porcelain and indeed all semiconductors and insulators, as well as electrolytes, show a decrease in resistance with rise in temperature.

Resistivity and temperature coefficient of resistance

Substance		Resistivity (in Ω m)		Temp. coefficient of resistance (per K)	
		Value	Temp. (K)	Value $\times 10^{-4}$	Temp. range (K)
Metals	Copper, drawn	$1·78 \times 10^{-8}$	291	42·8	70–670
	Aluminium	$3·21 \times 10^{-8}$	291	38	70–770
	Iron, pure	$11·5 \times 10^{-8}$	323	62	70–570
	Platinum	$11·0 \times 10^{-8}$	291	38	70–1500
	Eureka or constantan	$49·0 \times 10^{-8}$	291	$-0·4$ to $+0.1$	—
	Manganin	$44·5 \times 10^{-8}$	291	0·02–0·5	—
	German silver	$16–40 \times 10^{-8}$	291	2·3–6	—
	Platinoid	$34·4 \times 10^{-8}$	291	2·5	—
Semi-conductors	Germanium	0·47	291	All	—
	Silicon	3000	291	negative with a	—
Insulators	Glass	$10^{10}–10^{11}$	291	non-linear	—
	Diamond	10^{14}	291	relation-	—
	Mica	9×10^{14}	291	ship	—

The platinum resistance thermometer

The variation in the resistance of platinum with temperature is utilized in the platinum resistance thermometer.

A fine platinum wire is wound in a double spiral on a mica former enclosed in a tube of fused silica and connected by relatively thick copper leads to one arm of a Wheatstone bridge (see Fig. 2.23). If the

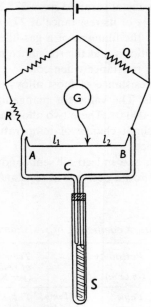

Fig. 2.23

resistance of the platinum wire is of the order of 2 or 3 Ω at 273 K, it increases by about 1 Ω for a rise of 100 K. Thus to measure temperature to $\frac{1}{100}$ K, it is necessary to measure a change of resistance of $\frac{1}{10\,000}$ Ω. This fine adjustment is achieved by means of a sliding contact and a wire, AB, of suitable uniform resistance per unit length, r Ω per cm. The resistances of the arms P and Q are made equal and R is made approximately equal to the resistance of the platinum wire S.

When the bridges is balanced, since $P = Q$,

$$R + rl_1 = S + rl_2.$$
$$\therefore \ S = R + r(l_1 - l_2).$$

The effect of a change in resistance with temperature of the copper leads to the platinum wire is eliminated by means of similar, compensating leads C.

The resistance of platinum varies with temperature, as measured on the perfect gas scale, in accordance with the equation

$$R_t = R_0(1 + \alpha t + \beta t^2),$$

where R_t and R_0 are the resistances at t and 273 K respectively, and α and

β are constants. To find the three unknown constants R_0, α and β it is necessary to determine the resistance of the platinum wire at three fixed points; the boiling-point of sulphur under standard atmospheric pressure (717·60 K) is used in addition to the usual ice and steam.

The advantages of the platinum resistance thermometer are its accuracy and its wide range, 70 to 1400 K.

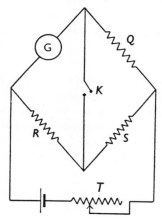

Fig. 2.24

The resistance of a galvanometer

The most accurate method of finding the resistance of a galvanometer is to clamp the moving part and insert the galvanometer in the arm of a Wheatstone bridge. It is not always possible, however, to do this, in which case the method of Fig. 2.24, due to Lord Kelvin, may be employed. The variable resistance T is adjusted until a suitable deflection is obtained in the galvanometer. The resistances R, S or Q are adjusted until, on depressing the switch K, there is no change in the deflection of the galvanometer. The potentials of the corners of the bridge connected through K must then be at the same potential, and the bridge is balanced. Hence

$$G/Q = R/S.$$

Measurement of low resistance by a potentiometer

The Wheatstone bridge is not suitable for measuring resistance of less than $\frac{1}{10}$ Ω, since the resistances of the connecting wires and at the terminal contacts are appreciable. Low resistances are most conveniently

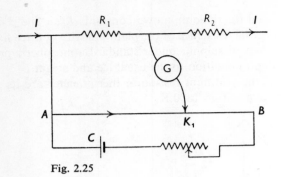

Fig. 2.25

measured by means of a potentiometer. A suitable current, I, is passed through the unknown low resistance, R_1, and a standard low resistance, R_2, in series (see Fig. 2.25). The p.d.'s across the two resistances, IR_1 and IR_2, are compared by means of a potentiometer. When the balance point, K_1, has been found, the leads from A and from the sliding contact are connected to the ends of R_2 (instead of R_1) and a new balance point, K_2, is found.

$$\frac{IR_1}{IR_2} = \frac{AK_1}{AK_2}.$$

$$\therefore \frac{R_1}{R_2} = \frac{AK_1}{AK_2}.$$

Suppose the resistances of R_1 and R_2 are of the order of 0·1 Ω and I is 1 A; then the p.d.'s across R_1 and R_2 are about 0·1 V. If the lead acid accumulator, C, were connected directly across AB the p.d. across AB would be about 2 V and K_1 would be very near to A. The variable resistance is adjusted until the balance point, K_1, is a reasonable distance from A: the p.d. across AB is then probably of the order of 0·2 V and the rest of the p.d., 1·8 V, is across the variable resistance.

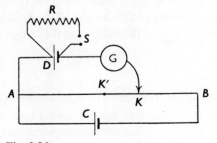

Fig. 2.26

Measurement of internal resistance of a cell by a potentiometer

The internal resistance of a cell may be measured by comparing the p.d.'s between the terminals of the cell when on open circuit and when bridged by a known resistance, R (Fig. 2.26). The balance point K on the potentiometer wire is determined with the switch, S, open, and the balance point, K', with the switch closed.

Let E = e.m.f. of cell, i.e. the p.d. on open circuit.

        V = p.d. of cell when connected to R.

        I = current flowing through R.

        r = internal resistance of cell.

By Ohm's law

$$E = I(R+r) \quad \text{and} \quad V = IR.$$

$$\therefore \frac{R+r}{R} = \frac{E}{V} = \frac{AK}{AK'}.$$

Knowing R, AK and AK' it is possible to calculate r.

Measurement of a high resistance

The Wheatstone bridge method is unsuitable for the measurement of high resistances – of the order of 1 MΩ.

To measure a high resistance the method of substitution can be employed. The resistance is placed in series with a battery and sensitive galvanometer whose deflection is noted. The unknown resistance is replaced by a standard high resistance which is adjusted until the deflection is the same as before. The value of the standard high resistance is equal to that of the unknown resistance.

An alternative method is a direct application of Ohm's law; the high resistance is placed in series with a galvanometer of known sensitivity and a suitable known p.d. is applied.

The ohmmeter

A galvanometer can be converted into an ammeter by means of a shunt and into a voltmeter by means of a high series resistance. It is possible also to use a galvanometer as an ohm-meter; its scale may be graduated in ohms.

The galvanometer is shunted with a variable resistance S and is connected in series with a cell, usually of e.m.f. 1·5 V, and a standard resistance, R_0 (Fig. 2.27). The scale is graduated 'backwards'; a full-scale deflection reads zero ohms.

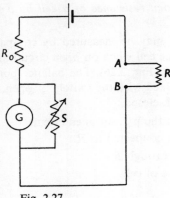

Fig. 2.27

Before using the instrument the terminals AB are short-circuited and the shunt resistance, S, is adjusted to give a full-scale deflection, i.e. a zero reading. S is necessary to compensate for the change of the e.m.f. of the battery with age.

Suppose the current is I_0 when the terminals are short-circuited, and suppose it is I when an unknown resistance R is put across the terminals. If E is the e.m.f. of the battery, ignoring its internal resistance and the resistance of the galvanometer plus shunt,

$$E = I_0R_0 = I(R+R_0).$$

$$\therefore \quad R = R_0\left(\frac{I_0}{I} - 1\right).$$

Since R_0 and I_0 are constants, the value of R depends on the current I, and hence the scale can be graduated in ohms. The mid-scale reading, when $I = \frac{1}{2}I_0$, is clearly equal to R_0.

A universal meter, with scales graduated in amperes, volts and ohms, consists of a moving-coil galvanometer, a set of shunts, a set of series resistances, a battery and an elaborate switch to enable the appropriate connections to be made, all enclosed in a case.

Summary

Current may be measured with a galvanometer, or voltameter, or potentiometer in conjunction with a standard resistance and a standard cell.

Potential difference may be measured by means of a high-resistance galvanometer, i.e. a voltmeter, or by means of a potentiometer.

Resistance may be measured by a Wheatstone bridge, or by a potentiometer (low resistance), or by the method of substitution (high resistance), or by an ohmmeter.

The resistivity of a substance is given by:

$$\rho = \frac{RA}{l}.$$

and equals R when l and A are unity.

The temperature coefficient of resistance of a metal is defined by the equation

$$R_t = R_0(1+\alpha t).$$

Questions

1. (a) Explain the principle of the moving-coil galvanometer.

(b) Describe how you would find the current sensitivity of a galvanometer of this type, giving a diagram of your electrical circuit.

(c) A milliammeter of resistance 6 Ω reads up to 10 mA. How would you convert it into an ammeter reading to 3 A? (N)

2. A certain galvanometer gives a full-scale deflection for $\frac{1}{2}$ mA, and has a resistance of 20 Ω. How can it be adapted for use as (a) an ammeter to measure currents up to 5 A, (b) a voltmeter to measure potentials up to 5 V?
(O & C)

3. A galvanometer whose resistance is 12 Ω gives full-scale deflection for $\frac{1}{100}$ A. What modifications would be necessary to make it read (a) voltages up to 100 V, (b) currents up to 100 A? (OS)

4. How may a milliammeter of resistance 20 Ω be adapted to act as a voltmeter, so that the scale formerly reading milliamperes now reads the same number of volts? (OS)

5. In order to find the current sensitivity of a galvanometer it is shunted with a resistance of 10 Ω and connected in series with a cell and a resistance of 1 MΩ. If the galvanometer has a resistance of 90 Ω, the cell an e.m.f. of 1·5 V, and the deflection obtained is 150 scale divisions, find an approximate value for the sensitivity of the galvanometer in amperes per scale division. (N)

6. A simple moving-magnet galvanometer consists of a small magnet suspended by a fine silk thread at the centre of a coil of wire. Explain how it works and why it is more suitable for use with a potentiometer or Wheatstone bridge than for measuring currents. How can its sensitivity be increased with the aid of a bar magnet?

7. Explain the action of a moving-iron ammeter. What are its advantages and disadvantages as compared with a moving-coil instrument?

8. Give the theory of the potentiometer method of comparing the electromotive forces of two cells.

A student who has put together such a circuit fails to find a dead-point. State the possible reasons for this.

(Contd. over.)

A Daniell cell on open circuit gives a dead-point when the terminals are connected across 0·90 m of potentiometer wire. When the cell terminals are connected across a 4 Ω coil, the length of wire necessary for a balance is reduced to 0·80 m. Calculate the internal resistance of the cell.   (C)

9. A 2 V accumulator of negligible internal resistance is connected to the ends of a potentiometer wire 1 m long and 5 Ω resistance. A cell of e.m.f. 1·5 V and internal resistance 0·9 Ω is connected through a galvanometer in the usual way; what length of potentiometer wire will be required to produce a balance?

What length of potentiometer wire will be required to balance in the following cases:
  (a) When 1 Ω is placed in series with the accumulator?
  (b) When 1 Ω is placed in parallel with the cell of e.m.f. 1·5 V?   (L)

10. Explain the principle of the potentiometer.

You are supplied with the following apparatus: an accumulator of e.m.f. about 2 V; a standard cell of e.m.f. 1·018 V; a metre wire of resistance 1 Ω; two variable resistance boxes each providing resistances up to 2 kΩ; a sensitive galvanometer and connecting wire. Explain with the help of a labelled diagram how, using this apparatus, you would arrange and standardize a potentiometer which would measure potential differences up to 1 mV along the potentiometer wire.   (N)

11. In an experiment to determine the resistivity of copper, a length of copper wire of resistance r_1 is connected in series with a comparable standard resistance r_2 of 0·01 Ω, an accumulator of just over 2·00 V e.m.f. and a resistance r of 20 Ω. The potential differences across r_1 and r_2 are compared by a potentiometer arrangement consisting of a uniform wire of 30 Ω resistance, a resistance R in series, a 2 V accumulator, and the usual accessories. Draw the circuit diagram. What is the purpose of including the resistance r in the circuit? Estimate the approximate value of R if the balance points are near the middle of the potentiometer wire.   (N)

12. An accumulator, a resistance box, a coil of resistance 400 Ω, and a potentiometer wire of length 10 m and resistance 40 Ω, are connected in series. It is found that for a certain value of the resistance in the box, 6·85 m of potentiometer wire are required to balance the e.m.f. of a concentration cell. For the same value of the box resistance the potential drop across the 400 Ω resistance plus 3·20 m of potentiometer wire just balances a standard cell of e.m.f. 1·018 V. Calculate the e.m.f. of the concentration cell.   (O & C)

13. Describe the Wheatstone bridge method of measuring resistances, and prove the formula connecting the resistances when a balance is obtained.

In a Wheatstone bridge the four resistances in the arms of the bridge are AB 2 Ω, BC 4 Ω, AD 1 Ω, and DC 3 Ω. The terminals of a cell of e.m.f. 2 Ω and negligible resistance are connected by wires of negligible resistance to A and C. If a galvanometer of resistance 10 Ω is connected between B and D, find the current in the galvanometer.   (O & C)

14. Discuss in general terms the difficulties which present themselves in the measurement of low and of high resistance by the simple Wheatstone bridge method.

Describe methods, one for each case, which are suitable for measuring a resistance of order 0·1 Ω, and one of order 100 MΩ.   (N)

15. A telegraph line BCD, which is 50 km long, has a fault due to earthing at an unknown point C. The end B of the line is joined to the end D through resistances P, Q, R in series and in that order. A battery is connected from the end B to the junction of the resistances Q and R and a galvanometer is connected from the junction of P and Q to earth. The resistance of R is equal to that of 12·5 km of the telegraph line. When $P = 1500\ \Omega$ and $Q = 1425\ \Omega$ the galvanometer is not deflected. Where is the fault in the line?   (D)

16. What is the effect upon the balance of a Wheatstone bridge of interchanging the battery and the galvanometer?

17. How does the resistance of a material in the form of a wire vary (a) with its length, (b) with its area of cross-section? State briefly how you could verify these relations experimentally.

18. The cross-section of the live rail of an electric railway is 57 cm² and the resistivity of the iron is $11·5 \times 10^{-8}\ \Omega$m. Neglecting the effect of joints, calculate the resistance of 1 km of the rail.

19. Given that the resistivity of aluminium is twice that of copper, and that the density of aluminium is one-third that of copper, find the ratio of the masses of aluminium and copper conductors of equal length and equal resistance.   (N)

20. A wire 10 cm long tapers uniformly from a diameter of 1·0 mm at one end to a diameter of 0·50 mm at the other end. If the resistivity of the material of the wire is $20 \times 10^{-8}\ \Omega$ m, what is the resistance of the wire?

21. Describe an experiment to determine the temperature coefficient of resistance of copper.

A metal wire, 100 cm long and of 10^{-6} m² cross-section, has a resistance of 0·20 Ω at 273 K. Calculate the specific resistance of metal at 773 K given that its temperature coefficient of resistance is 62×10^{-4} per K, and neglecting its thermal expansion.   (N)

22. The resistance of a length of wire is 1·054 Ω at 293 K, and its temperature coefficient of resistance is 0·003 93 per K. Calculate its resistance (a) at the minimum winter temperature of 263 K and (b) at the maximum summer temperature of 303 K.

23. The resistance of an electric lamp is found to be 55 Ω when measured by a Post Office box at 288 K. When it is connected to the 230-V supply it uses 100 W. Assuming that the resistance of the lamp is proportional to the absolute temperature, find the working filament temperature.   (B)

24. Give an account of the variation of the electric resistance of metals with temperature.

Describe how the variation of the resistance of a platinum wire may be applied to determine the temperature of a bath of hot oil.   (OS)

25. The resistance of the platinum wire of a platinum resistance thermometer increases by 4·0 Ω for a rise in temperature of 100 K. If the bridge wire (AB in Fig. 2.23 is 50·0 cm long, find what its resistance must be in order that a movement

of 0·5 mm of the sliding contact may correspond to a change of temperature of 0·01 K.

26. (a) Describe the Wheatstone bridge method of comparing resistances. Discuss whether, in the version known as the metre bridge, it is necessary, unnecessary or desirable that (i) the slide wire be of uniform resistance per unit length, (ii) the galvanometer be very sensitive and (iii) the e.m.f. of the battery be constant.

(b) Distinguish between *electromotive force* and *potential difference*. In what circumstances can the potential difference between the terminals of a battery exceed its e.m.f.?

A battery of e.m.f. 8 V and internal resistance 30 Ω is used to drive a current through a resistance of 50 Ω. A voltmeter whose resistance is 250 Ω is placed across the 50 Ω resistor. What does the voltmeter read?   (O & C 1969)

27. Draw a diagram of a potentiometer circuit suitable for measuring the e.m.f. of a cell known to be about 1 V.

A potentiometer is sometimes described as a voltmeter whose resistance is infinite. Explain this statement and discuss the extent to which it is justified in practice.

Explain how a potentiometer could be used to find the ratio of two resistances each known to be about 0·1 Ω. Draw a diagram of a suitable circuit and indicate on it appropriate values for the components.

The e.m.f. of a cell measured using a potentiometer is found to be 1·5 V. A voltmeter connected across the terminals of the cell reads 1·25 V. Explain the discrepancy and calculate the ratio of the resistance of the voltmeter to the internal resistance of the cell.   (O & C 1971)

28. Given a cell of e.m.f. 1·0 V as a reference standard, how would you measure accurately the e.m.f. of

(a) a thermocouple of e.m.f. about 5 mV;

and (b) a power pack of e.m.f. about 5 kV and internal resistance about 1 MΩ?

In each case give approximate values for the components in your circuit and estimate (with justification) the accuracy that you would expect to achieve.

(O & C 1970)

29. A uniform wire, of length 2 m and cross-sectional area 10^{-5} m², carries a steady current of 1·04 A. The resistivity of the material is $4·8 \times 10^{-6}$ Ω m. What is the potential drop down each centimetre of the wire?

What current would be needed to give a potential drop of 5 mV down the whole length of the wire, and what resistance would be needed in series with the wire and a 2 V driver cell (of negligible resistance) to get this current?

Describe, with full experimental details, how you would use a potentiometer to determine the e.m.f. of a thermocouple when the two junctions are maintained at two steady fixed temperatures.   (O 1969)

30. State Kirchhoff's laws for currents flowing in a network of conductors.

'The first law is simply the law of conservation of charge; the second law is the law of conservation of energy.' Discuss this statement briefly.

Figure 2.28 shows an unbalanced Wheatstone's bridge connected across a battery of e.m.f. 2 V and negligible internal resistance. Calculate (a) the p.d. between the points A and B, (b) the current passing through the galvanometer G.

(O 1968)

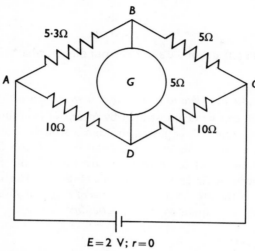

$E = 2$ V; $r = 0$

Fig. 2.28

3   Magnetic flux

An electric current sets up, in the space surrounding it, what is known as a *magnetic field*.

Magnetic field due to a current in a straight wire

The magnetic field can be demonstrated by means of iron filings. Suppose a current of about 5 A is passed through a straight vertical wire, inserted through a small hole in a horizontal card. If iron filings are sprinkled on the card and the card is gently tapped, the filings set themselves in concentric circles round the wire (Fig. 3.1). The lines along which the filings set themselves are called *lines of magnetic flux*. A number of lines represents a quantity of *magnetic flux*.

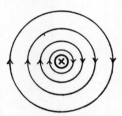

Fig. 3.1. Magnetic field due to a current in a straight wire passing down into the paper.

A small magnetic compass needle tends to set itself tangentially to a line of magnetic flux and hence at right angles to a straight wire carrying a current. The positive direction of a line of magnetic flux is taken as the direction in which the north pole of a compass needle tends to point. This is indicated by arrowheads in Fig. 3.1, in which the current is passing down into the paper, represented by $\otimes$. A current passing up out of the paper is represented by $\odot$. The positive direction of the lines of magnetic flux due to a current in a wire can be remembered by means of *Maxwell's corkscrew rule: imagine a corkscrew screwed along the wire in the direction of the current; the direction in which the thumb rotates is the positive direction of the lines of magnetic flux.*

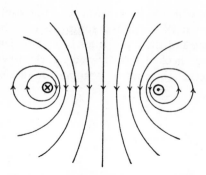

Fig. 3.2. Magnetic field due to a current in a circular coil. The current is passing down into the paper on the left and up out of the paper on the right.

Magnetic field due to a current in a circular coil

The magnetic field due to a current in a circular coil is shown in Fig. 3.2. Note that the magnetic field at the centre of the coil is at right angles to the plane of the coil. The current is passing down into the paper on the left and up out of the paper on the right.

Magnetic field due to a current in a solenoid

A coil of wire in the shape of a spiral, i.e. one that can be made by wrapping wire round a cylinder, is called a *solenoid*. The magnetic field due to a current in a solenoid is shown in Fig. 3.3. In the middle of the solenoid the lines of magnetic flux are parallel to the axis, but near the ends of the solenoid, known as its poles, they diverge from the axis.

The lines of magnetic flux outside the solenoid are similar to those of

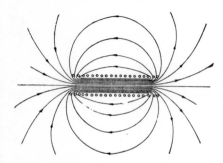

Fig. 3.3

Fig. 3.4

a bar magnet; a solenoid, which is freely suspended horizontally in the earth's magnetic field, sets itself north and south. Its ends behave like the poles of a bar magnet and their polarities can be remembered by Fig. 3.4, which should be checked by Maxwell's corkscrew rule. The polarities depend on whether the current is going round clockwise or counter-clockwise.

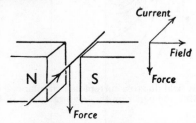

Fig. 3.5. Force on a wire in a magnetic field due to a current in the wire.

The force on a wire in a magnetic field

A wire carrying a current at right angles to a magnetic field experiences a force at right angles to itself and to the field. This can be demonstrated by passing a current through a wire held between the pole pieces of a powerful electromagnet as in Fig. 3.5.

The positive direction of the magnetic field between the pole-pieces is from the north pole-piece to the south pole-piece; the north pole of a compass needle is attracted towards the south pole-piece and repelled away from the north pole-piece.

The direction of the force on the wire can be predicted by means of *Fleming's left-hand rule*. Hold the thumb and first two fingers of the left hand mutually at right angles (Fig. 3.6). Then if the *f*irst finger

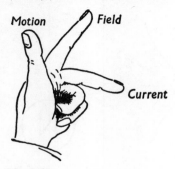

Fig. 3.6

points in the direction of the *f*ield, the se*c*ond finger in the direction of the *c*urrent, the thu*m*b will point in the direction of *m*otion (force).

If the wire is turned so that it is no longer at right angles to the magnetic field, the force on it is reduced. When the wire lies along the direction of the magnetic field it experiences no force. This provides an alternative method of defining the direction of the magnetic field (instead of the direction in which a compass needle would point).

The force can be explained as due to the interaction of the magnetic field due to the current and of the external magnetic field (Fig. 3.7).

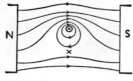

Fig. 3.7

Faraday imagined that the lines of magnetic flux are in tension and tend to shorten, thereby providing a kind of mechanical model to account for the force on the wire.

These qualitative ideas may be made precise by experiments performed on a 'force-on-a-conductor balance'. Firstly, the force on the conductor, F (in newtons), is measured for various values of the current I (in amperes), whilst the length l (in metres) is kept constant. It is found that

$$F \propto I.$$

Next F is measured for various values of l, whilst I is kept constant, it is found that

$$F \propto l$$

so $$F \propto Il.$$

During both these experiments the third possible variable, the magnetic field, is kept constant.

Writing the constant of proportionality as B, the two experiments above suggest

$$F = BIl.$$

B will clearly vary with the magnetic field.

Magnetic flux density

The previous equation $F = BIl$ is used to define B. B is a measure of the strength of the magnetic field and is called the 'magnetic flux density'.

WILHELM WEBER

JOSEPH HENRY

WILHELM WEBER (1804–1890) *was foremost among the founders of the system of electrical measurements. He made absolute measurements of newly devised concepts such as current, potential difference, resistance, capacitance, and inductance and showed how these measurements could be interlinked in a comprehensive system. In the course of his investigations he invented several well-known instruments, such as the earth inductor and the dynamometer, and discovered that the velocity of light is involved in a system of electrical units. He was called to the chair of physics at Göttingen in 1831, where he worked with Gauss, a pioneer in magnetic measurements. He based the absolute measurement of current on the force exerted by a current on a unit magnetic pole, as defined by Gauss; but this has been replaced by the force between parallel currents. Resentment was aroused in Germany when, at the congress in Paris in 1881 at which the names of the electrical units were selected, Gauss and Weber, the founders of the system adopted, were overlooked, and some German writers used 'weber' instead of 'ampere' to denote the unit of electric current. The term 'weber' is now used internationally, however, as the unit of magnetic flux.*

JOSEPH HENRY (1797–1878) *discovered electromagnetic induction independently of Faraday, but published his work later; he did, however, gain priority in the discovery of self-induction and this is commemorated in the unit of inductance named after him. He taught at the Albany Academy, New York State, became professor of natural philosophy at Princeton University in 1882, and for the rest of his long life, was the first director of the Smithsonian Institution, Washington, founded by the bequest of an Englishman. The Smithsonian Institution was the parent body of several of the national scientific institutions of the USA, such as the Meteorological Bureau and the National Museum. While there, Henry's work was primarily that of scientific administration, in which he displayed great ability, particularly in the difficult times of the Civil War. It has been said that he would have done more for the advancement of American science had he devoted his undoubted genius more exclusively to research and obtained more recognition of his work. His researches in electromagnetic induction developed from his improvement of the electromagnet. He demonstrated the principle of electric telegraphy, constructed a primitive electric motor, and made the discovery basic to the work of Hertz, that the discharge of a Leyden jar is oscillatory.*

Thus the magnetic flux density is defined as the force per unit length experienced by a conductor carrying unit current perpendicular to the field. It is clear that B is a vector quantity. Writing

$$B = \frac{F}{Il}$$

and substituting F in newtons, I in amperes, and l in metres gives us the units of B.

$$\frac{F}{Il} = \frac{N}{A \times m} = \frac{N\,m}{A\,m^2} = \frac{J}{A\,m^2}$$

$$= \frac{A\,V\,s}{A\,m^2} = \frac{V\,s}{m^2}.$$

A 'Vs' is called a 'weber' (pronounced 'vayber') and is abbreviated to Wb. So the unit of magnetic flux density B is Wb m^{-2}; this unit has been named tesla and given the symbol T.

If the wire is not at right angles to the magnetic field but is at an angle θ to the field, then the component of the magnetic flux density at right angles to the wire must be used and the formula for the force on the wire becomes $BIl \sin \theta$.

EXAMPLE

What is the force exerted on a straight wire of length 3·5 cm, carrying a current of 5 A, and situated at right angles to a magnetic field of flux density 0·2 Wb m^{-2}?

Force $= BIl$
$= 0·2 \times 3·5 \times 10^{-2} \times 5$
$= 0·035$ N.

Force on an electron moving in a magnetic field

An electric current in a wire is conventionally regarded as a flow of positive charge, although it consists in fact of a flow of negative electrons in the opposite direction.

Suppose an electron of charge e is moving with velocity v at right angles to a magnetic field of flux density B.

The electron moves a distance l in a time t, where $t = l/v$, and constitutes a current I.

Current $=$ flow of charge per second.

$$\therefore I = \frac{e}{t} = \frac{e}{l/v} = \frac{ev}{l}$$

$$\therefore Il = ev.$$

But force on a current $= BIl$.
$\therefore$ Force on a moving electron $= Bev$.

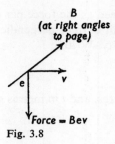

Fig. 3.8

Figure 3.8 represents the direction of the force on the electron which should be verified by Fleming's left-hand rule, remembering that the direction of the positive current is opposite to the direction of motion of the electron.

Torque on a rectangular coil

Figure 3.9 (*a*) and (*b*) represents a vertical rectangular coil of length and breadth *a* and *b* respectively, carrying a current *I* with its plane at an

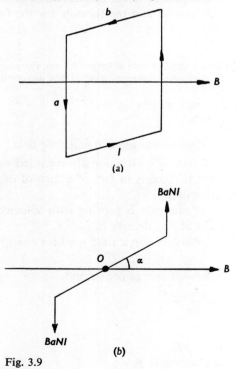

Fig. 3.9

angle α to a horizontal magnetic field of magnetic flux density B. Applying Fleming's left-hand rule to Fig. 3.9 (*a*), it will be seen that the left-hand vertical side is urged out of the paper, the right-hand vertical side into the paper, and the top and bottom are urged up and down respectively. If the coil is free to turn about a vertical axis, only the forces on its vertical sides will have a turning effect. The forces on these sides are each $BaNI$, where N is the number of turns of the coil; they are shown in Fig. 3.9 (*b*).

$$\text{Moment of the forces about } O = 2BaNI\frac{b}{2}\cos\alpha$$

$$= BANI\cos\alpha \text{ newton-metre.}$$

where A = area of coil = ab. Thus the torque on the coil is $BANI\cos\alpha$ and its maximum value, when the plane of the coil is parallel to B and $\alpha = 0$, is $BANI$. Its minimum value, when the plane of the coil is perpendicular to B and $\alpha = 90°$, is zero.

It can be shown that the torque on the coil is always $BANI\cos\alpha$ whatever the shape of the coil.

Moving-coil galvanometer

The coil of a moving-coil galvanometer will be in equilibrium when the torque on it due to the current is equal to the torque of the controlling forces, provided by the torsion of a suspending phosphor-bronze fibre or by hair-springs. The controlling torque is proportional to the angle turned through. If we assume that, when no current flowed, the coil was lying with its plane in the direction of the magnetic field, and that the controlling torque per radian is c newton-metre,

$$\text{Controlling couple} = c\alpha$$

$$\therefore \ BANI\cos\alpha = c\alpha$$

$$I = \frac{c}{BAN}\frac{\alpha}{\cos\alpha}.$$

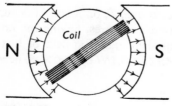

Fig. 3.10

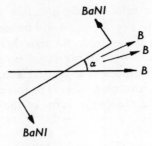

Fig. 3.11

In order that the current I may be proportional to the deflection α, the field is made radial by means of a soft-iron cylinder (Fig. 3.10). The vertical sides of the coil are then always in a magnetic field which is in the direction of the plane of the coil, whatever the position of the coil, and hence the forces on the sides of the coil are always at right angles to its plane (Fig. 3.11). The torque is now $BANI$ instead of $BANI \cos \alpha$. Thus

$$I = \frac{c}{BAN} \alpha.$$

The sensitivity of a galvanometer is the deflection per unit current.

$$\text{Sensitivity} = \frac{\alpha}{I} = \frac{BAN}{c}.$$

EXAMPLE

Find the torque on a galvanometer coil, 2 cm square and containing 100 turns, when a current of 1 mA flows through it. The radial field of the permanent magnet has a flux density of $0 \cdot 2$ Wb m^{-2}.

$$\begin{aligned}
\text{Torque} &= BANI \\
&= 0 \cdot 2 \times 2^2 \times 10^{-4} \times 100 \times 1 \times 10^{-3} \\
&= 8 \times 10^{-6} \text{ N m.}
\end{aligned}$$

Electromagnetic moment of a coil

It was shown on p. 55 that the torque, T, on a coil of N turns, area A square metres, carrying a current I amperes, having the normal to its plane at an angle θ with a uniform magnetic field of flux density B webers per square metre (Fig. 3.12) is $BANI \cos (90° - \theta)$ newton-metre, i.e.:

$$T = BANI \sin \theta.$$

This has a maximum value of $BANI$ when $\theta = 90°$, and the plane of

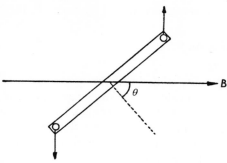

Fig. 3.12

the coil lies along the direction of the field. The coil tends to set itself with its plane perpendicular to the field.

The maximum torque, T_m, divided by the magnetic flux density, B, of the field is called the *electromagnetic moment*, M, of the coil. Thus

$$M = \frac{T_m}{B}$$

and is measured in A m².

The magnetic moment is a vector whose direction is taken as perpendicular to the plane of the coil; we shall have no need to distinguish between positive and negative directions.

The Biot–Savart law

We have seen that a current flowing in a wire causes a magnetic field and we have defined a measure of the magnetic field, the magnetic flux density B. Next we wish to calculate the field due to currents flowing in circuits of different shapes.

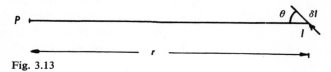

Fig. 3.13

We take as our starting point an expression for the magnetic flux density of a current I in an extremely short length δl of a circuit; $I\delta l$ is known as a 'current element'. Referring to Fig. 3.13, the magnetic flux density, δB, at P due to the current element is perpendicular to the paper and up out of the paper (corkscrew rule, p. 48), i.e. at right angles to the plane containing the current element and the point P; its magnitude is given by

$$\delta B = \frac{\mu I \, \delta l \sin \theta}{4\pi r^2}$$

where μ is the permeability of the medium.

This expression is often called the Biot–Savart law. It is not capable of direct experimental proof but is justified by the many verifiable deductions which can be made from it. We shall use it to deduce the magnetic flux density due to a number of simple circuits, at certain points for which the calculation is easy.

Permeability

The Biot–Savart law cannot actually be arrived at experimentally because it only holds in the limit of a vanishingly small current element. Rather the equation is *integrated*, as we shall do next, and then it may be checked. However we could in our imagination perform an experiment in which we hold all the likely variables but one constant and observe how the 'element of field, δB' is affected by that one variable. In this way we imagine ourselves to find δB proportional to I, but inversely proportional to r^2 etc. The constant of proportionality is found to vary depending on the medium in which we perform our imaginary experiments. It is written $\mu/4\pi$ in our system of units and μ is called the permeability of the medium. Its units are (V s/A m) or H m^{-1} (see p.103).

B at the centre of a circular coil

The value of B due to a circular coil is shown in Fig. 3.2 (p. 49). It is clear that the field is not uniform and the point for which the calculation of B is simplest is the centre of the coil, where the field is at right angles to the coil. Suppose that the coil has a radius of a and N turns, and that a current of I flows in it. Dividing the coil into current elements, each current element is at the same distance, a, from the centre of the coil and is at right angles to the line joining it to the centre. The magnetic flux density, B, at the centre is the sum of the terms δB for all the current elements.

$$B = \sum \frac{\mu I \, \delta l \sin \theta}{4\pi r^2}.$$

But $\sum \delta l = 2\pi a N$ and $\sin \theta = 1$, since $\theta = 90°$.

$$\therefore B = \frac{\mu 2\pi a N I}{4\pi a^2} = \frac{\mu N I}{2a}.$$

B at a point on the axis of a circular coil

Consider a circular coil of radius a carrying a current I (Fig. 3.14). The magnetic flux density, δB, at the point P on its axis, due to a current element $I\,\delta l$, will be $I\,\delta l \sin \theta / 4\pi r^2$, where $\theta = 90°$ and $\sin \theta = 1$. The direction of δB is at right angles to the line joining P to the current

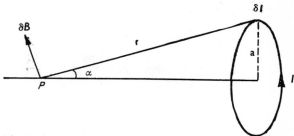

Fig. 3.14

element and in the plane of the paper (the current element being perpendicular to the paper); δB can be resolved into a component along the axis, $(I\,\delta l / 4\pi r^2) \sin \alpha$ and one at right angles. By considering pairs of current elements at opposite ends of a diameter it is clear that the magnetic flux density at right angles to the axis vanishes.

If the coil has N turns, $\sum I\,\delta l$ for the whole coil becomes $I2\pi Na$. Magnetic flux density along the axis at P due to the whole coil,

$$B = \frac{\mu I 2\pi Na}{4\pi r^2} \sin \alpha$$

$$= \frac{\mu NI}{2a} \sin^3 \alpha \quad \left(\frac{a}{r} = \sin \alpha \right).$$

B on the axis of a solenoid

Consider a point P on the axis of a solenoid carrying a current I having n turns per metre and of radius a (Fig. 3.15). To find the value of B at P the solenoid may be divided into a large number of short lengths, such as LM, and each of these regarded as a circular coil. LM subtends the small angle $\delta\alpha$ at P.

$$LM = \frac{r\,\delta\alpha}{\sin \alpha} \quad \text{(see Fig. 3.15 (b)).}$$

The number of turns in a length LM is $(nr\,\delta\alpha / \sin \alpha)$. Applying the expres-

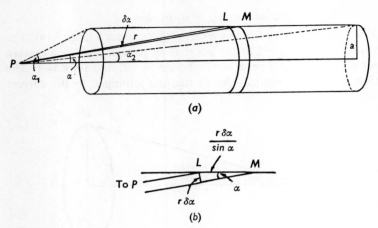

Fig. 3.15

sion for the magnetic flux density on the axis of a circular coil to obtain the magnetic flux density, δB, at P due to the short length of solenoid LM,

$$\delta B = \frac{(\mu n r \, \delta\alpha/\sin \alpha)I}{2a} \sin^3 \alpha.$$

But $$r = \frac{a}{\sin \alpha}$$

$$\therefore \; \delta B = \tfrac{1}{2}\mu n I \sin \alpha \, \delta\alpha.$$

$\therefore$ Magnetic flux density at P due to the whole solenoid

$$= \int_{\alpha_2}^{\alpha_1} \tfrac{1}{2}\mu n I \sin \alpha \, d\alpha$$

$$= \tfrac{1}{2}\mu n I (\cos \alpha_2 - \cos \alpha_1).$$

If the solenoid is infinitely long (when P lies inside it)

$$\alpha_1 = \pi \quad \text{and} \quad \alpha_2 = 0.$$

$$\therefore \; B = \tfrac{1}{2}\mu n I (\cos 0 - \cos \pi)$$

$$= \mu n I.$$

If P is at one end of a long solenoid, $\alpha = \tfrac{1}{2}\pi$ and $\alpha_2 = 0$.

$$\therefore \; B = \tfrac{1}{2}\mu n I (\cos 0 - \cos \tfrac{1}{2}\pi)$$

$$= \tfrac{1}{2}\mu n I.$$

Thus the value of B at one end of a long solenoid is half that in the middle of the solenoid.

Minimum length of a 'long' solenoid

We can now calculate the error involved by taking the magnetic flux density at the middle of a solenoid of finite length as μnI. Suppose that the length of the solenoid is l and its diameter of cross-section is d. Referring to Fig. 3.15 and imagining P to lie at the centre of the solenoid,

$$\cos \alpha_2 = \frac{\frac{1}{2}l}{\sqrt{\{(\frac{1}{2}l)^2+(\frac{1}{2}d)^2\}}} = \frac{1}{\sqrt{(1+d^2/l^2)}},$$

$$\cos \alpha_1 = \frac{-\frac{1}{2}l}{\sqrt{\{(\frac{1}{2}l)^2+(\frac{1}{2}d)^2\}}} = -\frac{1}{\sqrt{(1+d^2/l^2)}}.$$

Magnetic flux density at centre of solenoid

$$= \tfrac{1}{2}\mu nI(\cos \alpha_2 - \cos \alpha_1)$$

$$= \mu nI \frac{1}{\sqrt{(1+d^2/l^2)}}$$

$$= \mu nI(1+d^2/l^2)^{-1/2}$$

$$= \mu nI\left(1-\frac{1}{2}\frac{d^2}{l^2}\right)$$

(neglecting terms of higher power since they are small compared with d^2/l^2).

Thus the value of B is less than that at the centre of an infinitely long solenoid by the fraction $\frac{1}{2}(d^2/l^2)$. If $d = \frac{1}{10}l$ this fraction is $\frac{1}{200}$ and the error involved is $\frac{1}{2}\%$. For an error of 1%,

$$\frac{1}{2}\frac{d^2}{l^2} = \frac{1}{100},$$

$$\frac{d^2}{l^2} = \frac{1}{50},$$

$$d \simeq \tfrac{1}{7}l.$$

B at a given distance from a long straight wire

To find the value of B due to a current I in a long straight wire at a point P distant a from the wire, consider the current element $I\,\delta x$ in Fig. 3.16,

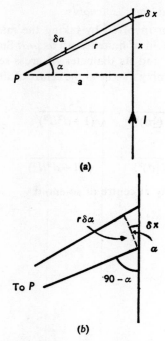

Fig. 3.16

which produces a δB at P at right angles to the paper, up out of the paper.

$$\delta B = \frac{\mu I \, \delta x \sin (90-\alpha)}{4\pi r^2},$$

$$\delta x = \frac{r \, \delta \alpha}{\cos \alpha} \quad \text{(see Fig. 3.16 } (b))$$

$$= \frac{a \, \delta \alpha}{\cos^2 \alpha} \quad \left(\text{since } r = \frac{a}{\cos \alpha} \right).$$

$$\therefore \ \delta B = \frac{(\mu I a \, \delta \alpha / \cos^2 \alpha) \cos \alpha}{4\pi (a^2 / \cos^2 \alpha)}$$

$$= \frac{\mu I \cos \alpha \, \delta \alpha}{4\pi a}.$$

To find the value of B at P due to the *whole* wire we must integrate the expression for δB.

$$B = \int_{-\pi/2}^{\pi/2} \frac{\mu I \cos \alpha \, d\alpha}{4\pi a}$$

$$= \frac{\mu I}{4\pi a} \left[\sin \alpha\right]_{-\pi/2}^{\pi/2}$$

$$= \frac{\mu I}{2\pi a}.$$

EXAMPLE

Find the magnetic flux density, B, at the centre of a square coil, of side $2a$, carrying a current I (Fig. 3.17).

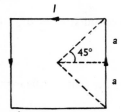

Fig. 3.17

Value of B due to one side of the coil

$$= \int_{-\pi/4}^{\pi/4} \frac{\mu I \cos \alpha \, d\alpha}{4\pi a} \quad \text{(see above)}$$

$$= \frac{\mu I}{4\pi a} \left[\sin \alpha\right]_{-\pi/4}^{\pi/4}$$

$$= \frac{\sqrt{(2\mu I)}}{4\pi a}.$$

$\therefore$ Magnetic flux density, B, due to the four sides $= \dfrac{\sqrt{2}\mu I}{\pi a}$.

Force between two long straight wires

Two parallel currents flowing in the same direction attract one another. This can be understood from the combined magnetic field (Fig. 3.18), assuming the lines of magnetic flux to be in tension and to repel each other laterally. It can also be deduced from Fleming's left-hand rule.

Two parallel currents flowing in opposite directions repel one another.

Suppose two long parallel wires carrying currents I_1 and I_2 are at a distance a apart.

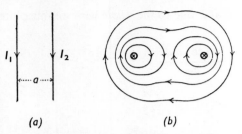

(a)   (b)

Fig. 3.18. Attraction of two like currents.

magnetic flux density B due to first wire at second wire

$$= \frac{\mu I_1}{2\pi a} \quad \text{(p. 63)}.$$

Force per metre exerted on second wire $= BI_2$ (p. 51)

$$= \frac{\mu I_1 I_2}{2\pi a} \text{ newtons per metre}.$$

The units of μ are $\text{VsA}^{-1}\text{m}^{-1}$, and hence the units of the expression for the force per metre are

$$\frac{\text{V s}}{\text{A m}} \times \frac{\text{A}^2}{\text{m}} = \frac{\text{J}}{\text{m}^2} = \frac{\text{N}}{\text{m}}.$$

The first wire experiences an equal force in the opposite direction.

Permeability of a vacuum

From the definition of the ampere (p. 2) the force per metre on two long parallel wires, each carrying a current of 1 A, and 1 m apart, is 2×10^{-7} N. Hence if μ_0 is the permeability of a vacuum,

$$2 \times 10^{-7} = \frac{\mu_0 \times 1 \times 1}{2\pi \times 1}.$$

$$\therefore \mu_0 = 4\pi \times 10^{-7} \frac{\text{weber}}{\text{ampere metre}} \text{ (or henry metre}^{-1})$$

$$= 4\pi \times 10^{-7} \text{ H m}^{-1}.$$

Thus the value of μ_0 can be deduced from the definition of the ampere.

To determine a system of electrical units, one independent electrical unit must be defined in addition to the mechanical units. In the SI system the ampere is regarded as this independent electrical unit.

Measurement of current by a dynamometer

Dynamometer instruments are similar in principle to a moving-coil galvanometer of which the permanent magnet is replaced by a fixed coil. The moving coil and the fixed coil are put in series and hence each carry the same current I. The torque is proportional to I^2 and does not change in direction when the direction of the current is changed. A dynamometer instrument is therefore particularly suitable for measuring alternating currents and is in common use in the form of a wattmeter (p. 212).

We shall describe a method of measuring direct current with a dynamometer, however, because this provides a simple method for the absolute measurement of current, as an alternative to the current balance described on p. 68.

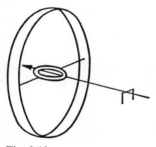

Fig. 3.19

A small circular coil is pivoted on a horizontal axis at the centre of a large, fixed, vertical, circular coil (Fig. 3.19) and the coils are connected in series. When a current is passed through the coils the pivoted coil tends to turn until its plane is at right angles to the field of the large coil, i.e. until it is vertical, so that its magnetic field reinforces the magnetic field of the large coil. This can be checked by regarding the pivoted coil as a short bar magnet (see Fig. 3.4, p. 49), setting itself in the field of the large coil.

The pivoted coil is retained in a horizontal position by moving a rider along a horizontal pointer attached to it. The torque on the coil is then equal to the torque due to the rider. The plane of the large coil is set in the magnetic meridian to eliminate the torque due to the earth's magnetic field.

Let the current be I and the number of turns and the radii of the large and small coils be N_1 and N_2, and a_1 and a_2 respectively.

Magnetic flux density at centre of large coil

$$= \frac{\mu_0 N_1 I}{2a_1} \quad \text{(p. 58)}.$$

Torque on small coil $= BAN_2 I$ (p. 55)

$$= \frac{\mu_0 N_1 I}{2a_1} \pi a_2^2 N_2 I$$

$$= \frac{\mu_0 N_1 N_2 \pi a_2^2 I^2}{2a_1}.$$

Let the mass of the rider be m, x be the distance of the rider from the axis of the pivoted coil, and g be the acceleration due to gravity.

Torque due to rider $= mgx$.

Equating the two torques,

$$I^2 = \frac{2mgxa_1}{\mu_0 N_1 N_2 \pi a_2^2}.$$

All the quantities on the right-hand side of the equation are known, or can be measured, and hence I can be obtained. The units of the right-hand side are

$$\frac{\text{newton} \times \text{metre} \times \text{metre}}{\text{henry metre}^{-1} \times \text{metre}^2} = \frac{\text{N} \times \text{m} \times \text{A}}{\text{V s}}$$

$$= \frac{\text{J} \times \text{A}}{\text{J A}^{-1}} = \text{A}^2.$$

EXAMPLE

The fixed, vertical coil of a dynamometer has a radius of 15·0 cm and 120 turns; the pivoted coil has a radius of 2·00 cm and 100 turns. A rider of mass 0·050 g must be moved a distance of 13·0 cm from the axis of the pivoted coil, to keep the plane of the latter horizontal, when a current is passed through the coils. Calculate the current.

$$I^2 = \frac{2mgxa_1}{\mu_0 N_1 N_2 \pi a_2^2}.$$

$mg = 0·05 \times 10^{-3} \times 9·81$ N, $x = 0·13$ m, $a_1 = 0·15$ m, $a_2 = 0·02$ m,

$N_1 = 120$, $N_2 = 100$, $\mu_0 = 4\pi \times 10^{-7}$ H m^{-1}.

$$\therefore I^2 = \frac{2.0·05.10^{-3}.9·81.0·13.0·15}{4\pi.10^{-7}.120.100.\pi.0·02^2}.$$

$I = 1·01$ A.

Absolute measurements

The electrical measurements made in the normal laboratory are based on standard resistances and standard cells. It was explained in Chapter 2, for example, how the calibration of an ammeter, or of a voltmeter, can be checked using these two standards. The two standards are calibrated by means of absolute measurements, i.e. measurements based on the fundamental definitions of the units, for which responsibility is undertaken by the national physical laboratories of the world, such as the National Physical Laboratory at Teddington near London, and the Bureau of Standards at Washington, USA. The two absolute measurements involved are those of resistance and current. Indeed, the whole system of electrical and magnetic standards could be based on these two absolute measurements.

All electrical and magnetic units can be derived from mechanical units, plus one additional electrical or magnetic unit. The additional unit selected is the ampere, which we defined on p. 2. The definition of the ampere implies taking μ_0, the permeability of free space, as $4\pi \times 10^{-7}$ henry per metre. Alternatively we could have taken the fourth fundamental unit, in addition to those of mass length and time, as μ_0.

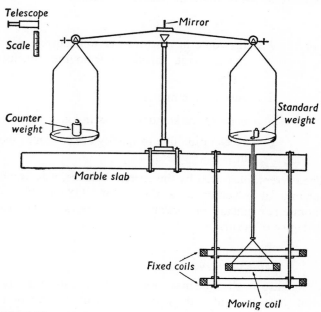

Fig. 3.20

Absolute measurement of the ampere by current balance

The absolute measurement of the ampere is made by means of a current balance and the result is expressed in terms of the mass of silver deposited in electrolysis. One of several very accurate determinations is that of Curtis and Curtis made in 1934 at the Bureau of Standards, Washington.

The current balance (Fig. 3.20) consisted of a movable coil arranged between two larger fixed coils, the force between the fixed and moving coils being measured by a balance capable of weighing to 1 part in 100 million.

The standard mass was first removed and a current sent through the coils connected in series in such a way that the moving coil was forced downwards. The counterweight was adjusted so that the balance swung about its position of rest, the swings being observed by a telescope and scale.

The current in the fixed coils was then reversed causing the moving coil to be forced upwards and the standard mass replaced. The standard mass was of such a value that, for the particular current being measured, it was equal to twice the force between the coils, thus causing the balance once more to swing about its position of rest. Then

$$2FI^2 = mg,$$

where $F =$ the force in newtons per unit current exerted on the moving coil, calculated from the dimensions of the coils,

$I =$ current in amperes,

$m =$ mass of the standard mass in kg,

$g =$ acceleration due to gravity at the pan of the balance in $m\ s^{-2}$.

The current, measured by the current balance, was found to deposit silver in a silver voltameter at the rate of $1{\cdot}118\ 04 \times 10^{-6}\ kg\ A^{-1}\ s^{-1}$. Elaborate refinements and precautions are necessary to measure to this high degree of accuracy.

Absolute measurement of the volt

Once we have an absolute measurement of the ampere, a way to make an absolute measurement of the volt follows from its definition as the watt per ampere. A known current is passed through a resistive circuit and the heat energy output measured. Then if V is the voltage across the circuit and I the absolute value of the current

VI = rate of heat output.

In the next chapter we show how an absolute determination of resistance can be made using the laws of electromagnetic induction. An absolute determination of the volt can then be made using Ohm's law

$V = IR.$

Summary

The magnetic flux density, B, measures the strength of a magnetic field and is defined by the equation:

$F = BIl.$

Its units are Wb m^{-2}, or T.

The force on an electron moving at right angles to a magnetic field

$= Bev.$

The electromagnetic moment of a current carrying coil (or a bar magnet) is defined by

$$M = \frac{T_m}{B}$$

where T_m is the maximum torque experienced in the field.

The magnetic field due to a current element is

$$\delta B = \frac{\mu I \, \delta l \sin \theta}{4 \pi r^2}.$$

This is called the Biot–Savart Law. Integrated this law yields:

B at the centre of a long solenoid $= \mu n I$

B at the centre of a circular coil $= \dfrac{\mu N I}{2a}$

B on the axis of a cirular coil $= \dfrac{\mu N I}{2a} \sin^3 \alpha$

B on the axis of a solenoid $= \frac{1}{2}\mu n I(\cos \alpha_2 - \cos \alpha_1)$

B at a distance, a, from a long straight wire $= \dfrac{\mu I}{2\pi a}.$

The force between two long straight parallel wires is then

$$F = \frac{\mu I_1 I_2}{2\pi a}.$$

From the definition of the ampere in terms of a force between two long straight wires, the permeability of the vacuum, μ_0, is

$$\mu_0 = 4\pi \times 10^{-7} \text{ H m}^{-2}.$$

The permeability of an arbitrary medium may be defined in terms of the force between two wires in the medium (e.g. $\mu = 2\pi aF/I_1 I_2$) or using any of the other integrated forms of the Biot–Savart law.

Questions

1. Two long solenoids are wrapped with the same length of wire; they are of equal axial length but one has a diameter of cross-section twice that of the other. Compare the magnetic flux densities at their centres when they carry the same current.

2. The magnetic flux density in the middle of a long solenoid carrying a current of $2 \cdot 0$ A is $5 \cdot 0$ mWb m^{-2}. Find the number of turns per metre of the solenoid. ($\mu_0 = 4\pi \times 10^{-7}$ H m^{-1}.)

3. State the formula for the magnetic flux density at any point due to a current element and apply it to find the magnetic flux density at the centre of a circular coil.
   Find the magnetic flux density at the centre of a circular coil of 2 turns, with a of $7 \cdot 0$ cm and carrying a current of $1 \cdot 0$ A. ($\mu_0 = 4\pi \times 10^{-7}$ H m^{-1}.)

4. Calculate the value of the magnetic flux density, B, at the centre of a circular coil of 50 turns of wire of radius 7 cm, carrying a current of $0 \cdot 5$ A; and also at positions on the axis of the coil distant $3 \cdot 5$, 7 and 14 cm from the centre of the coil. Plot a graph of B against distance from the centre of the coil. ($\mu_0 = 4\pi \times 10^{-7}$ H m^{-1}.)

5. Deduce an expression for the magnetic flux density on the axis of a solenoid of finite length.
   A solenoid, of length 40 cm and diameter of cross-section $4 \cdot 0$ cm, contains 800 turns and carries a current of $0 \cdot 5$ A. Calculate and plot the magnetic flux density on the axis of the solenoid (a) at the centre, (b) at $10 \cdot 0$ cm from the centre, (c) at one end, (d) at $10 \cdot 0$ cm beyond one end. ($\mu_0 = 4\pi \times 10^{-7}$ H m^{-1}.)

6. Calculate the magnetic flux density at the mid-point of the axis of a solenoid 20 cm long and of diameter of cross-section 10 cm, consisting of 600 turns, when carrying a current of $2 \cdot 0$ A. What is the magnetic flux density at the centre of a long solenoid having the same number of turns per metre and carrying the same current? ($\mu_0 = 4\pi \times 10^{-7}$ H m^{-1}.)

7. What is the magnetic flux density midway between two long parallel wires separated by 10 cm, each carrying a current of 5 A (a) in the same direction, (b) in opposite directions?
   What is the force per cm length exerted by each wire on the other? ($\mu_0 = 4\pi \times 10^{-7}$ H m^{-1}.)

8. How would you show by experiment that the force on a straight conductor carrying a current in a magnetic field is perpendicular to the conductor and to the

direction of the magnetic flux density? Deduce that it is $BI \sin \theta$ per unit length, where B is the magnetic flux density, I is the current and θ is the angle between B and I.

A straight rod of mass 1 g cm^{-1} is hung vertically and can turn about its upper end. A current of 10 A is sent through it. Find how much it is deflected from the vertical in a place where the earth's horizontal flux density is $1 \cdot 8 \times 10^{-5}$ Wb m^{-2}.

(OS adapted)

9. A long vertical wire in which a current is flowing produces a neutral point with the earth's magnetic field (i.e. a point at which the resultant magnetic field is zero) at a distance of $5 \cdot 0$ cm from the wire. Calculate the current, assuming that the horizontal component of the earth's magnetic flux is $1 \cdot 8 \times 10^{-5}$ weber m^{-2}. (Take $\mu_0 = 4\pi \times 10^{-7}$ H m^{-1}.)

Draw a diagram of the lines of magnetic flux, showing the direction of magnetic north and the position of the neutral point.

10. Calculate the force of repulsion per unit length between two long parallel bus bars 20 cm apart when, as a result of a short-circuit, the current in each is 4000 A. ($\mu_0 = 4\pi \times 10^{-7}$ H m^{-1}.)

11. Define the ampere and show that the definition implies a value of $4\pi \times 10^{-7}$ H m^{-1} for the permeability of free space.

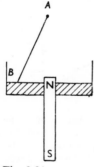

Fig. 3.21

12. In Fig. 3.21 AB is a light copper wire, freely pivoted at A and dipping at B into a trough of mercury, through the middle of which the N pole of a bar magnet protrudes. Describe and explain what occurs when a strong current is passed through the wire from A to B. What is the source of the energy for the occurrence?

13. Two parallel current elements, $I_1 \, \delta l_1$ and $I_2 \, \delta l_2$, are a distance r apart in a vacuum and the line joining their centres makes an angle θ with each. Write down an expression for the force which each exerts on the other, taking $\mu_0 = 4\pi \times 10^{-7}$ H m^{-1}.

14. A circular coil of 30 cm diameter and of 100 turns is fixed in a vertical plane and carries a current of 15 A. At its centre is hung, in the same vertical plane, a small square coil of 5-mm sides and of 160 turns. What forces act on the sides of this coil, and what is the torque exerted upon it, when a current of 1 mA is passed round it?

15. Describe how an absolute measurement of current may be made with a dynamometer.

A dynamometer consists of a fixed vertical coil, of radius 20·0 cm and 200 turns, and a smaller coil at its centre, of radius 2·5 cm and 100 turns, pivoted on a horizontal axis lying in the plane of the larger coil. A current of 1·50 A is passed through the two coils in series. Calculate the mechanical torque required to keep the plane of the smaller coil horizontal. ($\mu_0 = 4\pi \times 10^{-7}$ H m^{-1}.)

16. Show that the magnetic flux density B at the point P on the axis of the square coil in Fig. 3.22, having sides of length $2a$ and carrying a current I, is given by

$$H = \frac{\mu 2I}{\pi a} \sin \alpha \sin^2 \beta.$$

Hence show that the magnetic flux density, due to two such coils arranged co-axially and at a distance a apart at the mid-point between them is $\dfrac{32\mu I}{15\pi a}$ Wb m^{-2}.

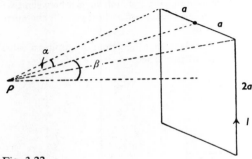

Fig. 3.22

17. Two similar, parallel, circular coils are mounted on the same axis but are separated by a distance equal to their radius (an arrangement due to Helmholtz). Obtain an expression for the magnetic flux density at a point on the axis midway between the coils, when they carry a current of I in the same direction. Take the number of turns of each coil as N and the radii as a.

If $a = 0.20$ metres by what percentage does the magnetic flux density at an axial point 1 cm from the mid-point differ from that at the mid-point?

18. Define *magnetic moment* of a coil and obtain an expression for it in terms of the current, the number of turns and the area of cross-section of the coil.

A pivoted coil, 2·5 cm square and of 50 turns, is situated at the centre of a long solenoid, with its plane parallel to the axis of the solenoid. The solenoid has 10 turns per cm and a current of 1·5 A is passed through both the coil and the solenoid. Calculate the torque on the coil. ($\mu_0 = 4\pi \times 10^{-7}$ H m^{-1}.)

19. A current I passes through a plane coil of n turns, each of area A. The plane of the coil makes an angle θ with a uniform magnetic field of flux density B. Obtain from first principles an expression for the couple experienced by the coil.

Explain the application of this result to the theory of the moving coil galvanometer.

A moving coil meter which has been calibrated at 288 K is used with an external circuit consisting of a constant-voltage source and copper conductors. For the copper coil of the meter, the temperature coefficient of resistance is $+4 \times 10^{-3}$ K^{-1}; for the control spring, the temperature coefficient (i.e. the fractional change in torque per degree rise in temperature) is -3×10^{-4} K^{-1}. Will the meter read too high or too low at temperatures above 288 K? Is the error significant at 298 K?

If the meter is to be converted to read as an ammeter, which would be the better material for the necessary shunt – one with a negligible temperature coefficient of resistance, or one with the same temperature coefficient of resistance as the coil?

(O 1968)

20 (*a*) Derive an expression for the current sensitivity, i.e. the deflection per unit current, of a moving coil galvanometer. Use this expression to discuss the problems posed by the design of a galvanometer of extremely high current sensitivity.

(*b*) Explain what is meant by an *absolute measurement* of an electrical quantity and explain the importance of such measurements. Describe an experiment to measure electrical resistance in absolute units.   (O & C 1969)

21. Draw a clear labelled diagram of a simple type of current balance. Explain the *physical principles* underlying the use of this instrument to measure currents. (You are not expected to deduce any formulae: it is sufficient to trace in qualitative terms the connection between the force which is measured and the current to be determined.) Why is such a balance just as effective for the measurement of alternating currents as it is for direct currents?

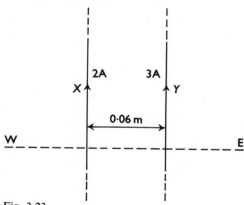

Fig. 3.23

Figure 3.23 represents two long parallel vertical wires 0·06 m apart. The current in *X* is 2 A and that in *Y* is 3 A, both flowing upwards.

(*a*) Considering only the effects of the currents flowing in them, calculate the force per metre experienced by each of the wires.

(*b*) Calculate the resultant force per metre on each wire when the earth's magnetic field is taken into account as well.   (*Contd. over.*)

(c) Why is knowledge of only the horizontal component of the earth's magnetic field required for the calculations in (b)?

(Take the value of μ_0 to be $4\pi \times 10^{-7}$ H m^{-1}, and the horizontal component of the flux density of the earth's magnetic field to be 2×10^{-5} T (i.e. Wb m^{-2}).)

(O 1970)

4   Electromagnetic induction

The magnetic field due to the current in a straight wire (see Fig. 3.1) becomes weaker with increasing distance from the wire. This may be represented by spacing the lines of magnetic flux farther apart at increasing distances from the wire. The magnetic field in the middle of a long, current-carrying solenoid is comparatively strong, and hence is represented by a considerable number of lines per unit cross-sectional area. It is also uniform, i.e. its strength is the same at every point, and hence it is represented by lines equally spaced. The uniformity of the magnetic field in the middle of a long solenoid is of fundamental importance in our development of the subject and we shall discuss an experimental proof on p. 85.

A complete line of magnetic flux is a closed loop; hence some of the lines in Figs. 3.2 and 3.3 are incomplete. Magnetic flux is said to be solenoidal. It does not begin or end anywhere but curls round an electric current.

The term flux is perhaps not a particularly happy one, since there is no question of flow or motion along the lines. Its origin lay in the analogy between lines of magnetic flux, whose property is to crowd together in regions where the magnetic field is strong and to spread apart where it is weak, and the lines of flow or vortex filaments in a swirling mass of fluid, which behave in a similar manner. Where the channel through which a fluid flows becomes narrower, the lines of flow crowd together and the fluid flow is faster; where the channel widens, the lines of flow spread out and the flow is slower. Again, lines of flow, like lines of magnetic flux, are closed loops because there is continuous circulation and no piling up of fluid. It is emphasized once more, however, that magnetic flux does not flow.

Electromagnetic induction

If a coil is wrapped round the outside of a solenoid carrying a current (Fig. 4.1), and the current in the solenoid is changed, an e.m.f. is induced in the coil; an induced current will flow in the galvanometer. The phenomenon is known as *electromagnetic induction*. The solenoid is

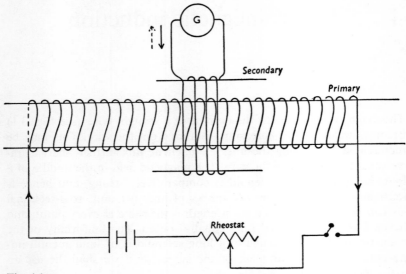

Fig. 4.1

called the *primary* and the coil is called the *secondary*. The induced
e.m.f. in the secondary lasts only while the current in the primary is
changing.

The phenomenon can be explained in terms of magnetic flux. The
magnetic flux set up by the current in the primary passes through the
secondary coil and is said to 'thread', or to be 'linked with', the
secondary coil. When the current in the primary is changed, the quantity
of magnetic flux threading the secondary changes, and this is regarded
as giving rise to the induced e.m.f. If the positions of any two circuits
are arranged so that all or some of the magnetic flux, set up by a current
in one of the circuits, threads the other circuit, then a change in the
current in one of the circuits gives rise to an induced e.m.f. in the other
circuit.

The direction of the induced current is such that it creates a magnetic
flux of its own in the opposite direction to that of the flux of the primary
when the flux of the primary is increasing, and in the same direction as
the flux of the primary when the flux of the primary is decreasing. In
this way the induced current tends to oppose the change which is causing
it. If it helped and augmented the change which caused it a perpetual
motion machine could be constructed, in violation of the principle of
the conservation of energy. In Fig. 4.1 the current in the secondary

circuit is in the direction of the continuous arrow when the current in the primary is switched off, or decreased; it is in the direction of the discontinuous arrow when the current in the primary is switched on, or increased.

The magnitude of the induced e.m.f. is proportional to the rate at which the magnetic flux is changed. Thus, in Fig. 4.1, if the current in the primary is changed slowly, by moving the slider of the rheostat slowly, the induced current, as recorded by the galvanometer, is smaller than if the change in the primary current were faster. On the other hand, the smaller induced current lasts for a longer time than the larger induced current, and the total quantity of electricity induced is the same for a given change of magnetic flux, whatever the rate of change of the magnetic flux.

The facts of electromagnetic induction can be summarized in two laws:

1. *Lenz's law*. The direction of the induced e.m.f. is such that it tends to oppose the change to which it is due.

2. *Faraday's law*. The magnitude of the induced e.m.f. is proportional to the rate of change of the magnetic flux to which it is due.

Unit of magnetic flux

The phenomenon of electromagnetic induction can be used to measure magnetic flux. The unit of magnetic flux is such that the rate of change of the magnetic flux is equal to the induced e.m.f. in volts.

Let us suppose that the flux changes at a uniform rate, from a value Φ_1 to a value Φ_2 in time t, so that the induced e.m.f., E, is constant during the whole time t. Let us further suppose that the secondary coil consists of a single turn so that we have to consider the induced e.m.f. in one turn only.

$$E = \text{rate of change of magnetic flux,}$$

i.e. $$E = -\frac{\Phi_2 - \Phi_1}{t}.$$

The minus sign indicates that the induced e.m.f. is in such a direction as to oppose the change of flux which gives rise to it. In a calculation of the magnitude only of the induced e.m.f., the negative sign can be disregarded.

Rewriting the above equation,

$$\Phi_2 - \Phi_1 = -Et.$$

This enables us to define change of magnetic flux as the product of the induced e.m.f. (in volts) and the time during which it lasts (in seconds). The unit of magnetic flux is the volt-second or *weber*.

Hence a change of magnetic flux of 1 weber, made at a uniform rate, produces an induced e.m.f. of 1 volt for 1 second; or, at a faster rate, of 10 volts for $\frac{1}{10}$ second, etc.

In practice, flux seldom changes at a uniform rate and, to represent the lack of uniformity, we must use the calculus notation. Suppose that a very small change of flux $\delta\Phi$, in a very short time δt, gives rise to a momentary induced e.m.f., E, at time t.

$$E = -\frac{\delta\Phi}{\delta t},$$

i.e.
$$E = -\frac{d\Phi}{dt}, \quad \text{when } \delta t \to 0.$$

The total change of flux, Φ, is the sum of all the small changes of flux $\delta\Phi$ during the whole time that the flux is changing:

$$\Phi = -\int E \, dt.$$

It is of interest to note the parallelism between the unit of electric charge and the unit of magnetic flux:

$$\text{coulomb} = \text{ampere-second},$$
$$\text{weber} = \text{volt-second}.$$

Magnetic flux linkage

A change in magnetic flux threading a coil causes an e.m.f. to be induced in each turn of the coil. If a magnetic flux Φ threads a coil of N turns, there is said to be a *magnetic flux linkage* of $N\Phi$ weber turns. When the flux changes uniformly

$$E = -N\frac{\Phi_2 - \Phi_1}{t}.$$

For a non-uniform change,

$$E = -N\frac{d\Phi}{dt}.$$

EXAMPLE

What is the e.m.f. induced in a coil of 50 turns when the magnetic flux threading through it changes at a uniform rate from 3 mWb to zero in $\frac{1}{100}$ s. ?

$$E = -N\frac{d\Phi}{dt} = -50 \cdot \frac{3 \times 10^{-3}}{1/100} = -15 \text{ V}.$$

Induced e.m.f. by cutting magnetic flux

If a wire is held between the poles of a powerful electromagnet, as in Fig. 4.2, and is moved sharply downwards, an induced e.m.f. is set up in the wire. The ends of the wire can be connected to a sensitive galvanometer, which will detect an induced current during the motion of the wire. The wire is said to 'cut' magnetic flux. 'Cutting' is essentially the same as 'threading', because the magnetic flux threading the secondary circuit, consisting of the wire and the galvanometer, changes during the motion of the wire; but cutting is a more convenient concept than threading when we are dealing with the induced e.m.f. in a single, straight conductor.

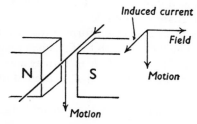

Fig. 4.2. Induced current in a wire due to its motion in a magnetic field.

The induced current flows only while the wire is moving. If the wire is moved quickly, a comparatively large current flows for a short time. If it is moved slowly a smaller current flows for a longer time. The total quantity of electricity induced is proportional to the total quantity of magnetic flux which is cut; the induced e.m.f. depends on the *rate* at which the magnetic flux is cut.

The direction of the induced current is such that it tends to oppose the motion of the wire (Lenz's law, p. 77). Thus the currents in Figs. 4.2 and 3.5 are in opposite directions because the force on the wire in Fig. 4.2, due to the interaction of the induced current and the magnetic field, must oppose the motion.

A right-hand rule, also due to Fleming, instead of a left-hand rule, is used to predict the direction of the induced e.m.f. The thumb and first two fingers of the right hand are held mutually at right angles. Then if the *f*irst finger points in the direction of the *f*ield, and the thu*m*b in the direction of *m*otion, the second finger will point in the direction of the induced *c*urrent or e.m.f.

In Chapter 3 we defined B from the empirical formula $F = BIl$. We

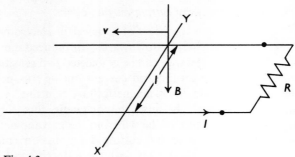

Fig. 4.3

will now show that if induced e.m.f. $= -d\Phi/dt$, and $\Phi = BA$, these two expressions containing B are compatible.

A frictionless rod XY is pushed with uniform velocity v through a uniform field B (Fig. 4.3). The induced e.m.f., E, causes a current I to flow through a resistance R. The current, I, in turn results in an opposing force, $-F$, so that an applied force, $+F$, is needed to maintain the steady velocity, v.

First we find the power needed to keep the rod moving, using the formula $F = BIl$.

$$\text{Power} = \frac{\text{work done}}{\text{time taken}}$$

$$= \frac{\text{force} \times \text{distance gone}}{\text{time taken}}$$

$$= \text{force} \times \text{velocity}$$

$$= BIlv.$$

Next, we find the power needed from the electrical expression EI.

$$\text{Power} = EI = -\frac{d\Phi}{dt}I = -\frac{d}{dt}(BA)I = BI \cdot -\frac{dA}{dt}.$$

But area swept in 1 second $= vl$, $\therefore dA/dt = -vl$,

$$\therefore \text{power} = BIlv, \text{ as before.}$$

Thus in a uniform field,

$$B = \frac{\Phi}{A}$$

where ϕ is the flux through an area A.

In a non-uniform field, the magnetic flux density at a point is the limiting value of the quotient when the flux $\delta\Phi$, through an imaginary small area around the point and perpendicular to the field direction, is divided by the area of the surface, δA.

$$B = \frac{d\Phi}{dA}$$

$$\Phi = \int B \, dA.$$

The unit of flux, the weber, is rather large, and quantities of magnetic flux met with in laboratory experiments are usually of the order of milliwebers (10^{-3} weber). On the other hand, magnetic flux densities of the order of 1 Wb m^{-2} (1 tesla) are not uncommon.

Quantity of electricity induced

We will now calculate the quantity of electricity, Q, induced in a coil of N turns and resistance R, by a change of flux, Φ. We will first assume that the flux changes uniformly from a value zero to a value Φ in a time t; the induced e.m.f., E, will be uniform during the change.

$$E = -N\frac{\Phi}{t}.$$

$$\therefore I = -\frac{N\Phi}{R\,t} \quad \text{(Ohm's law)}.$$

$$\therefore Q = It = -\frac{N\Phi}{R}.$$

We will now assume that the change of flux does not occur uniformly and use the calculus notation.

$$\text{Instantaneous induced e.m.f.} = -N\frac{d\Phi}{dt}.$$

$$\text{Instantaneous induced current, } I = -\frac{N}{R}\frac{d\Phi}{dt}$$

$$Q = \int I \, dt$$

$$= \int_{0}^{\Phi} -\frac{N}{R}\frac{d\Phi}{dt}\,dt$$

$$= -\frac{N\Phi}{R}.$$

The quantity of electricity induced is thus proportional to the change of the flux linkage, $N\Phi$, and inversely proportional to the resistance of the coil (or complete circuit including the coil), R.

EXAMPLE

What is the quantity of electricity induced in a coil of 50 turns, in a circuit of total resistance 250 Ω, due to a change of magnetic flux threading through the coil of 3 mWb?

$$Q = -\frac{N\Phi}{R}$$

$$= -\frac{50 \times 3 \times 10^{-3}}{250} = -6 \times 10^{-4} \text{ C.}$$

The ballistic galvanometer

It is clear from the equation $Q = -N\Phi/R$ that Φ can be determined by measuring Q.

A galvanometer specially designed to measure a quantity of electricity induced momentarily is known as a ballistic galvanometer. There are two important features of its design. The period of vibration of the moving part must be longer than that of a current-measuring galvanometer because the discharge must pass before the moving part has had time to leave its zero position. Also the galvanometer should be as far as possible undamped.

The coil of a *moving-coil, current* galvanometer is wound on a light metal frame to increase the damping. As the coil moves through the magnetic field of the permanent magnet, eddy currents in the frame and in the coil impede its motion. This makes it dead-beat, i.e. it takes up, without oscillation, its deflected position, which is independent of the damping, because the coil is then at rest and the current is still flowing. However, in the *ballistic* galvanometer, the discharge has ceased to flow long before the coil reaches its deflected position and damping reduces the throw.

The coil of a ballistic galvanometer should obviously not be wound on a metal frame. If the galvanometer is used to measure the discharge of a capacitor, no eddy currents can flow in the coil, because there is not a complete circuit, and hence there is no electromagnetic damping (there is always slight damping due to air resistance). But if the galvanometer is used to measure the quantity of electricity induced in a coil, the galvanometer circuit is completed through the coil and eddy currents flow. The magnitude of the eddy currents, and hence of the damping, will depend on the resistance of the galvanometer circuit. In the experiments we shall describe it will be arranged that the resistance of the galvanometer circuit is constant, so that the damping will affect all throws equally and can be neglected.

The maximum damping occurs when the galvanometer coil is short-circuited, because the eddy currents are then a maximum, and advantage may be taken of this to stop the coil from swinging to and fro, and to bring it quickly to rest ready for use.

Many galvanometers, not specifically designed for the purpose, can be used ballistically and most moving-coil milliammeters or micro-ammeters give reasonably satisfactory results.

Use of ballistic galvanometer to measure magnetic flux

To measure the flux density of a given magnetic field, the ballistic galvanometer is connected in series with a small coil of N turns, called a search coil. The coil is placed perpendicular to the field, when it is threaded by a flux Φ. Assuming that the field is uniform, and taking B as its flux density and A as the area of cross-section of the search coil, $\Phi = BA$. Suppose that R is the total resistance of the search coil and ballistic galvanometer circuit. The quantity of electricity, Q, induced when the search coil is jerked out of the field, is given by

$$Q = -\frac{N\Phi}{R}.$$

Thus the fling of the galvanometer, which is proportional to Q, is also proportional to the flux Φ. The scale of the galvanometer can therefore be calibrated direct in units of flux, i.e. webers, and the instrument composed of ballistic galvanometer and search coil is termed a flux-meter. A method of calibrating a ballistic galvanometer is described on p. 101, but all that we shall require in this chapter will be the ratio of the flings, when there is no need for calibration.

Magnetic flux density can also be measured using the Hall effect. (A voltage proportional to the field appears across a material carrying a current in the field.)

Investigation of the magnetic field of a solenoid

The magnetic field of a solenoid carrying a current can be investigated by means of a search coil connected to a ballistic galvanometer. A convenient method of finding how the flux density, B, varies along the axis of the solenoid is to wrap the search coil (of about 1000 turns) round a short length of a long cardboard tube which just slides *inside* the solenoid. In this way the search coil can be moved easily so that

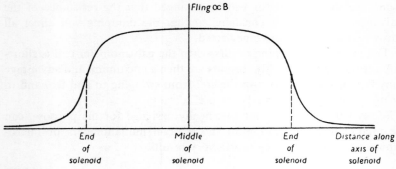

Fig. 4.4

its centre is at a number of points spaced along the axis of the solenoid. When the centre of the search coil is at each point, the current in the solenoid is switched off, with the result that a quantity of electricity, Q, is induced in the search coil and there is a fling in the galvanometer proportional to Q. Using the same nomenclature as above,

$$Q = - \frac{N\Phi}{R}$$

$$= - \frac{NBA}{R}.$$

The number of turns of the search coil N, the area of cross-section of the search coil A, and the total resistance of the galvanometer circuit R, remain constant as the search coil is moved. Hence $Q \propto B$, and each fling of the galvanometer is proportional to the flux density B at the position occupied by the search coil. The graph obtained by plotting galvanometer fling against distance along the axis of the solenoid is shown in Fig. 4.4.

The magnetic flux density at the ends of the solenoid is half that in the middle of the solenoid. Inside the solenoid, at distances from one end greater than about $3\frac{1}{2}$ times the diameter of cross-section, the flux density is 99 per cent of the calculated value for an infinitely long solenoid. Hence, in practice, a 'long' solenoid should have a length at least seven times its diameter (see also p. 61).

Uniform field in the middle of a long solenoid

If a search coil is placed near the middle of a long solenoid, but outside it, as in Fig. 4.5, the fling in the ballistic galvanometer is small when the current in the solenoid is switched off. Hence the flux density here is

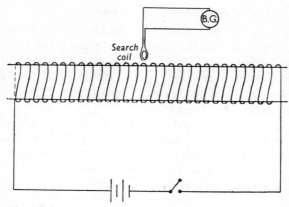

Fig. 4.5

small (less than 1 per cent of the maximum value inside the solenoid in the case of a solenoid whose length is seven times its diameter) and it is smaller the longer the solenoid, being zero for an infinitely long solenoid.

From this experimental fact we can deduce that the magnetic field in the middle of a long solenoid is uniform, i.e. that the flux density is constant at all points of the cross-section. Imagine four similar long solenoids of very small, square cross-section, with contiguous sides as shown in cross-section in Fig. 4.6, and carrying the same current. The magnetic fields inside the solenoids will be equal, since the solenoids are similar and the field in each is unaffected by its neighbours because their external fields are negligible. Currents in contiguous sides cancel out and the four solenoids can be replaced by a single solenoid, whose field in its four quarters will be the same. This process can be multiplied indefinitely so that we can regard a solenoid as made up of an infinite number of tiny solenoids each having the same internal field. Hence the field in the middle of a long solenoid is uniform.

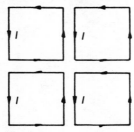

Fig. 4.6

Absolute measurement of the ohm

Two phenomena can be utilized to make an absolute measurement of resistance, either the heating effect of an electric current or electromagnetic induction.

The heat generated in a wire is I^2Rt (see p. 7). Since the absolute value of I can be found with a current balance the formula enables the absolute value of R to be calculated.

The best known determination of the absolute ohm by the electromagnetic method is that of Lorentz, later improved by Smith at the National Physical Laboratory. A copper disk has its axle coincident

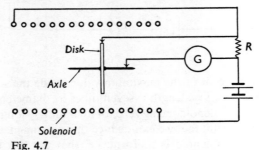

Fig. 4.7

with that of a long solenoid carrying a current (see Fig. 4.7). The disk is rotated at speed and the induced e.m.f. between the rim and the axle is balanced against the potential difference between the ends of the resistance to be determined, R, through which the same current flows as through the solenoid.

The induced e.m.f. is proportional to the rate at which the lines of magnetic flux are cut. Each radius of the disk cuts $\pi a^2 B/T$ lines of magnetic flux per second, where a is the radius of the disk, B the magnetic flux density inside the solenoid, and T the time for 1 revolution of the disk.

$$B = \mu_0 nI \quad \text{(p. 60)}.$$

$$\therefore \text{ Induced e.m.f.} = \frac{\pi a^2 \mu_0 nI}{T}.$$

P.d. across resistance $= IR$.

$$\therefore IR = \frac{\pi a^2 \cdot \mu_0 nI}{T},$$

$$R = \frac{\mu_0 \pi a^2 n}{T}.$$

Thus R is measured in terms of the radius of the disk a, the number of turns per unit length of the solenoid n, and the time of revolution of the disk T.

There are difficulties to be overcome such as the induced e.m.f. in the disk due to the earth's magnetic field; the smallness of the induced e.m.f. renders thermoelectric e.m.f.'s, due to heating at the rubbing contacts, important, and makes it necessary to send only a fraction of the current I through R. Moreover, the above simple calculation must be elaborated to allow for the lack of uniformity of the magnetic field inside the solenoid.

When the absolute value of one resistance has been found in this way it can be used to find the absolute values of other resistances by the Wheatstone Bridge method. Resistances can be compared with an accuracy of 1 in 10^6 fairly easily; the absolute ohm has been established to an accuracy of 1 in 10^5.

Mutual induction

A current changing in one circuit will induce an e.m.f. in a nearby circuit. Some of the magnetic flux due to the current in the first circuit, called the *primary*, must interlink the second circuit, called the *secondary* (Fig. 4.8). The phenomenon is called *mutual induction*.

The direction of the induced current in the secondary is such that it tends to oppose the change in the magnetic field of the primary. Thus when the current in the primary is growing, as in Fig. 4.8, the induced current in the secondary sets up an opposing magnetic field; when the current in the primary is dying away the induced current in the secondary sets up a supporting magnetic field in the same direction as that of the primary. The induced current in the secondary lasts only while that in the primary is changing.

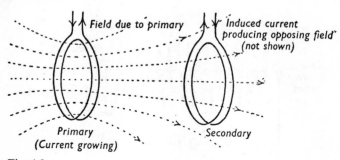

Fig. 4.8

Not all the magnetic flux from the primary in Fig. 4.8 links the secondary. The mutual induction could be increased to a maximum by moving the two coils together so that their faces were touching.

Self-induction

When a current in a coil (or indeed in any circuit) is changing, the magnetic flux links with the turns of the coil itself and gives rise to an induced e.m.f. The phenomenon is known as *self-induction*. When the current is increasing the induced e.m.f. opposes it, and when the current is decreasing the induced e.m.f. aids it.

Self-induction is of great importance in an a.c. circuit and it may be much more effective in reducing the current than the resistance of the circuit.

The induction coil

The induction coil (Fig. 4.9) is an instrument for producing a high voltage from a low one. A primary coil, consisting of a few thick turns of wire, is wound on a soft-iron core, and round it is wound a secondary coil consisting of many turns of fine wire. The current in the primary is switched on and off rapidly by means of a hammer make and break similar to that of an electric bell. The iron armature is attracted by the magnetized iron core and, as a result, the circuit is broken at the point of contact with the adjustable screw; the attraction then ceases and the springy steel causes the armature to fly back. This is repeated several times each second.

The magnetic flux due to the primary links with the turns of the secondary and by its variation induces currents in the secondary which are in such a direction that they set up a magnetic flux to oppose the change. If there are 1000 times as many turns in the secondary as in the

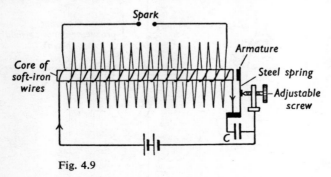

Fig. 4.9

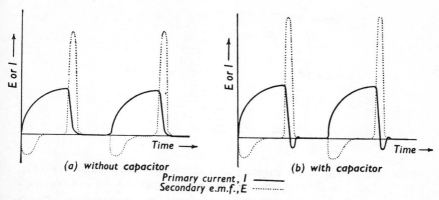

Fig. 4.10. (a) Without capacitor; (b) with capacitor. Primary current, I ———, secondary e.m.f., E · · · · · ·.

primary the induced e.m.f. in the secondary is approximately 1000 times the e.m.f. in the primary.

The magnitude of the secondary e.m.f. depends greatly on the rate at which the current changes in the primary. The self-induction in the primary causes the current, at make, to grow comparatively slowly; at break the primary current dies away more rapidly. The e.m.f. due to self-induction causes arcing between the adjustable screw and the contact on the springy steel, thus prolonging the duration of the primary current. This arcing can be reduced, and the induction coil made to work more efficiently, by means of the capacitor C which has a fairly large value, usually about 1 microfarad. Instead of jumping across the gap in the form of a spark the primary current charges the capacitor, and the latter then discharges through the primary coil, giving a small current in the opposite direction and causing an increase in the induced e.m.f. in the secondary. The effect of the capacitor is illustrated in Fig. 4.10. The secondary e.m.f. is much greater at the break than at the make of the primary current, and hence the induction coil acts almost as a unidirectional source of p.d.

Eddy currents, induced in the iron core, can be reduced by using a bundle of soft-iron wires, insulated from each other; the eddy currents tend to flow in circles round the axis of the core (as in the secondary coil), and the insulation between the iron wires blocks their flow.

The transformer

The transformer is an instrument for changing the magnitude of an alternating voltage and current. In its simplest form it consists of an

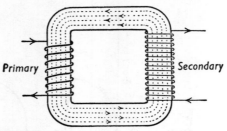

Fig. 4.11

iron ring round which primary and secondary coils are wound (Fig. 4.11). The dotted lines represent the magnetic flux due to the primary current, and, since this links with the secondary, a current will be induced in the secondary whenever the flux changes. The magnetic flux due to an alternating current in the primary is continually changing, so that an alternating e.m.f. of the same frequency is induced in the secondary. If the number of turns in the secondary is greater than the number in the primary, the voltage is *stepped up*; if it is less, the voltage is *stepped down*. Assuming that the flux leakage is negligible,

$$\frac{\text{voltage across secondary}}{\text{voltage across primary}} = \frac{\text{number of turns in secondary}}{\text{number of turns in primary}}.$$

The efficiency of transformers may be as high as 99 per cent. Hence it can be assumed, with reasonable accuracy, that

$$\text{power in primary} = \text{power in secondary.}$$

Thus if the voltage in the secondary is n times that in the primary, the current in the secondary is $1/n$ times that in the primary.

The current in the secondary sets up a magnetic flux of its own in a direction opposite to that of the primary when the primary current is increasing, and in the same direction when the primary current is decreasing. Hence an increase in the current drawn from the secondary causes a reduction in the back e.m.f. due to self-induction in the primary with a consequent rise in the primary current.

The ability of a transformer to change alternating voltages is of crucial importance in many areas, most notably in power transmission. It is also useful in electronics as a 'resistance changer'. If the primary and secondary alternating voltages are V_1, V_2 respectively, and the currents I_1, I_2 respectively, then

$$\frac{V_2}{V_1} = n$$

and $$\frac{I_2}{I_1} = \frac{1}{n}.$$

If the resistances associated with the primary, secondary are R_1, R_2 respectively, then

$$R_1 = \frac{V_1}{I_1}, \qquad R_2 = \frac{V_2}{I_2}$$

so $$\frac{R_2}{R_1} = \frac{V_2}{V_1} \times \frac{I_1}{I_2} = n \times n = n^2$$

and $$R_2 = n^2 R_1.$$

EXAMPLE

A valve should have a 16-kΩ load in its anode circuit but the loudspeaker it is required to feed is only 10Ω. What transformer would be required to match the loudspeaker to the valve?

Using $$R_2 = n^2 R_1$$

$$n = \sqrt{\left(\frac{R_2}{R_1}\right)} = \sqrt{\left(\frac{10}{16\,000}\right)} = \frac{1}{40}.$$

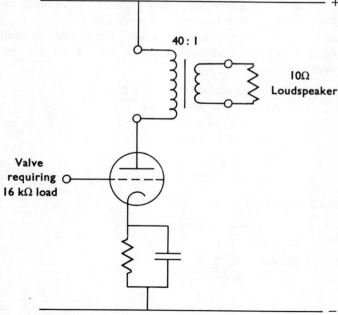

Fig. 4.12

So if a 40:1 step down transformer is used as in Fig. 4.12 the 10-Ω loudspeaker will appear to the valve to be a 16-kΩ load.

Eddy currents in the iron core are reduced by constructing the latter of thin, soft-iron sheets, called laminations, insulated from each others and of shape similar to the cross-section of the core shown in Fig. 4.11. The eddy currents tend to circulate in the same direction as the currents in the secondary, and hence are blocked by the insulation between the laminations.

The electric motor

The principle of the electric motor is illustrated in Fig. 4.13. A current is passed through a coil free to rotate between the two poles of a magnet. Applying Fleming's left-hand rule it will be seen that the coil in Fig. 4.13 tends to rotate in an anti-clockwise direction; the left-hand side of the coil is urged downwards and the right-hand side upwards. The rotation is made continuous by reversing the current each time the coil is vertical. This is done by means of a commutator, consisting of a metal ring split into two insulated halves connected to the ends of the coil, against which press carbon brushes. The current is led in and out through the brushes, and the latter change contact from one half-ring to the other half-ring, each time the coil is vertical.

A back e.m.f. is induced in the coil because the quantity of magnetic flux threading through it changes as it rotates. The motor, in this respect, behaves as a dynamo.

Let V = voltage applied across the coil,
        I = current through the coil,
        R = resistance of the coil,
        E_b = back e.m.f. in coil.

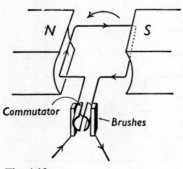

Fig. 4.13

Then $I = \dfrac{V - E_b}{R}$.

In a practical motor for which $V = 100$ volts, the back e.m.f., E_b, may be of the order of 95 volts. Multiplying the above equation by I and rearranging the terms,

$$VI \qquad = \qquad I^2R \qquad + \qquad E_bI$$

| Power supplied | Power dissipated as heat | Power to rotate coil |

From the principle of the conservation of energy, the energy expended by the source of supply in driving the current against the back e.m.f. is equal to the work done in rotating the coil.

If the rate of rotation of the coil if slowed down by making it do mechanical work (say against a brake), the back e.m.f. is reduced because the coil is cutting the lines of force more slowly, and hence the current through the coil increases. Thus an electric motor automatically regulates the amount of current and power required. A practical motor starting from rest must be connected in series with a resistance which is gradually reduced as the speed of the motor increases; otherwise owing to the absence of a back e.m.f., a very large, initial current will flow, and perhaps burn out the windings.

Practical motors

The torque or turning moment produced by the single coil in Fig. 4.13 is small and it fluctuates from a maximum when the coil is horizontal to zero when the coil is vertical. Hence to increase the torque in a practical motor, several coils are wound in slots in a soft-iron armature. Fig. 4.14 illustrates a drum-wound armature consisting of eight thick copper conductors interconnected as shown. The currents in the four conductors on the right-hand side are all passing into the paper,

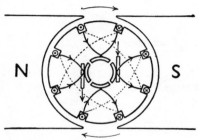

Fig. 4.14

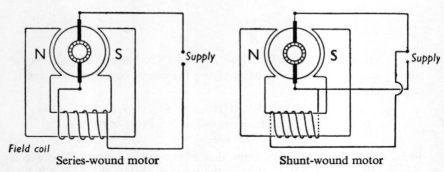

Field coil
          Series-wound motor   Shunt-wound motor

Fig. 4.15

and those on the left-hand side out of the paper. Thus all tend to rotate in a clockwise direction. Four segments are required on the commutator.

The air gap between the soft-iron armature and the poles of the field magnet is kept as small as possible in order to make the magnetic flux as large as possible. Eddy currents are reduced by constructing the armature of laminations.

The field magnet is usually an electromagnet and its coils may be connected either in series or in parallel with the armature coil (Fig. 4.15). *Series-wound* motors produce a large torque at starting, but the torque falls off with the load; parallel- or *shunt-wound* motors are not such good starters, but they are steadier with increasing load (see Fig. 4.16).

The torque of a motor is proportional to the product of the armature current and the magnetic flux, and hence to the product of the armature and field currents. The series motor is a good starter, because when there is a small back e.m.f., a large current flows in both the armature and field coils. In the case of the shunt-wound motor the current in the field coil is unaffected by the back e.m.f. in the armature coil, and only the armature current is large at starting.

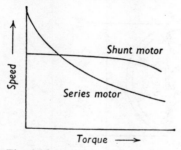

Fig. 4.16

Now consider the effect of increasing the load on the two motors. Each motor slows down and the back e.m.f. in the armature coil decreases, thus causing an increase in the armature current. In the case of the shunt-wound motor the current in the field coil is unaffected and the motor is able to provide the necessary increased torque at the decreased speed. But in the case of the series-wound motor the current in the field coils is also increased, causing a larger back e.m.f. in the armature coil, and hence reducing the current and causing a further drop in speed.

In order to obtain the advantages of both series- and shunt-wound motors, compound winding is employed; part of the field coil is in series and part in parallel with the armature coil.

The direction of rotation of a motor can be changed by reversing the current in either the armature or the field coil, but not in both. Thus a d.c. motor will run on a.c., but inefficiently, owing to the large self-induction of its coils.

The a.c. dynamo

If a coil is rotated by some external source of mechanical power between the poles of a magnet there will be generated in it an induced current which can be led away by slip rings (Fig. 4.17). The direction of the induced current in Fig. 4.17 can be verified by Fleming's right-hand rule (p. 79). The current is alternating, being a maximum when the coil is horizontal and zero when it is vertical.

The magnitude of the induced e.m.f. in the coil depends on the rate at which the coil is rotated, the area and number of turns of the coil, and also on the density of the magnetic flux. We will obtain an expression for the induced e.m.f. at any instant during the rotation.

Suppose the coil is plane, with N turns and of area A, and that it is rotated uniformly at a speed of n revolutions per second about an axis

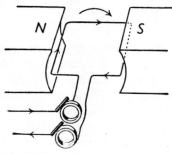

Fig. 4.17

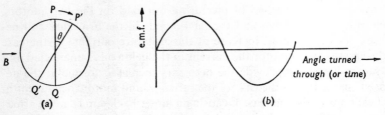

Fig. 4.18

at right angles to a uniform magnetic field of flux density B. Suppose the plane of the coil is initially vertical and perpendicular to the field, represented by PQ in Fig. 4.18 (a), and that after a time t it has turned through an angle θ and reached $P'Q'$.

Magnetic flux linkage through the coil in position $P'Q'$ = $BAN \cos \theta$.

∴ Induced e.m.f. in coil, $E = -\dfrac{d(BAN \cos \theta)}{dt}$

$$= BAN \sin \theta \frac{d\theta}{dt}.$$

But $\theta = 2\pi nt$ (since the coil turns through $2\pi n$ radians in 1 second)

$$\therefore \frac{d\theta}{dt} = 2\pi n,$$

$$\therefore E = BAN\,2\pi n \sin 2\pi nt.$$

If E_0 is the maximum or peak value of E, i.e. the value when the coil is parallel to the field and $\sin 2\pi ft = 1$,

$$E_0 = BAN\,2\pi n.$$

$$\therefore E = E_0 \sin 2\pi nt.$$

Also, using similar nomenclature for current,

$$I = I_0 \sin 2\pi nt.$$

Thus the induced e.m.f. (and also the current) may be represented by a sine curve as in Fig. 4.18 (b). The curve, known as the *wave form*, is only sinusoidal if the field and speed of rotation are uniform. In practice, the a.c. generated by turbo-alternators is very nearly sinusoidal. One complete wave, corresponding to one revolution of the coil, is called a cycle. The number of cycles per second, f, is called the *frequency* of the a.c. The angle measured along the horizontal axis in Fig. 4.18 (b) is called the *phase angle*.

EXAMPLE

A coil of 500 turns, of area 80 cm², is rotated at 1200 revolutions per minute about an axis at right angles to a magnetic field of flux density 0·25 Wb m⁻². Calculate (a) the maximum e.m.f. induced in the coil, (b) the e.m.f. when the coil is at 45° to the field, (c) the average value of the e.m.f.

(a) Maximum value of e.m.f.

$$= BAN \times 2\pi n$$
$$= 0 \cdot 25 \times 80 \times 10^{-4} \times 500 \times 2\pi \times \tfrac{1200}{60}$$
$$= 126 \text{ V.}$$

(b) Value of e.m.f. when coil is at 45° to the field

$$= BAN \times 2\pi n \sin 45°$$
$$= 126 \sin 45°$$
$$= 89 \text{ V.}$$

(c) The magnetic flux threads the coil in opposite directions every half revolution and hence changes $4n$ times per second.

$$\text{Average e.m.f.} = \text{change of flux per second}$$
$$= BAN \times 4n$$
$$= 0 \cdot 25 \times 80 \times 10^{-4} \times 500 \times 4 \times \tfrac{1200}{60}$$
$$= 80 \text{ V.}$$

Alternatively we can take the mean value of sin θ between 0 and π as $2/\pi$.

$$\therefore \text{ Average e.m.f.} = BAN \times 2\pi n \times 2/\pi$$
$$= BAN \times 4n.$$

The reverse motor effect of a dynamo

The induced current in a dynamo coil will tend to make it behave like a motor and to rotate in the opposite direction. The work which is done against the reverse motor effect, as it is called, is equal to the energy of the induced current. If the coil is connected to a high external resistance the induced current will be small and the reverse motor effect will likewise be small. If, however, the coil delivers a large current the reverse motor effect will be large and the coil will be harder to turn than before. When a dynamo is called upon to deliver a larger current, the torque provided, say by a steam engine, must be increased by a greater consumption of steam.

The d.c. dynamo

The current generated in the coil of a dynamo can be rectified by the use of a commutator as in Fig. 4.13 instead of slip rings. The current delivered to the external circuit will then be similar to Fig. 4.19. By the use of a drum-wound armature similar to Fig. 4.14, a current which fluctuates only slightly, as in Fig. 4.20, can be obtained.

The armature and field coils of a d.c. dynamo can be connected in

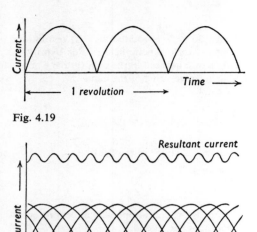

Fig. 4.19

Fig. 4.20

series or in parallel as in the case of a motor. Indeed Fig. 4.15 represents d.c. dynamos as well as motors. The voltage generated by a series-wound dynamo increases with the current it supplies while the voltage of a shunt-wound dynamo decreases (see Fig. 4.21). It is possible, by incorporating the two windings in a compound-wound dynamo, to keep the voltage approximately constant and independent of the load.

The voltage generated by a dynamo, as has already been stated, is proportional to the rate at which the armature coil cuts the magnetic flux. Since the speed of rotation of the armature coil is kept constant the voltage depends only on the current in the field coils. In the case of a series-wound dynamo an increase in the load (or current supplied) means an increase in the magnetic flux since the whole load current must pass through the field coil. Hence the voltage increases. In the

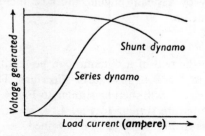

Fig. 4.21

case of a shunt-wound dynamo the field coil is in parallel with the external circuit, the e.m.f. being generated, of course, in the armature coil. A decrease in the resistance of the external circuit (i.e. an increase in the load) will cause this circuit to take a larger proportion of the current. Hence the voltage generated by a shunt-wound dynamo will fall when the load is increased owing to a decrease in the current in the field coil.

The telephone transmitter and receiver

Fig. 4.22 represents a telephone transmitter connected to a receiver. The transmitter contains carbon granules between two carbon blocks, one of which is rigidly attached to a diaphragm. Sound waves falling on the diaphragm cause it to vibrate and hence to subject the system to changes of pressure. Compression of the packing of the granules results in a comparatively large reduction in their electrical resistance. Thus the current in the line fluctuates in sympathy with the pressure changes in the sound waves.

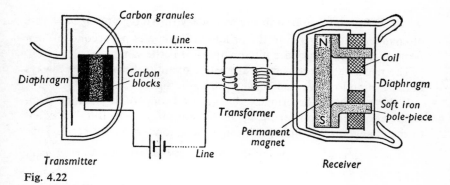

Fig. 4.22

The receiver consists of a permanent magnet, having two soft-iron pole-pieces wound with coils, and a steel diaphragm. When the fluctuating current set up by the transmitter passes through the coils of the receiver the force of attraction on the diaphragm fluctuates and hence the diaphragm vibrates, giving out sound waves corresponding to those which fall upon the transmitter. The purpose of the permanent magnet in the receiver is to maintain a continuous attraction on the diaphragm and thus to enable the diaphragm to respond more vigorously to changing currents in the coils. Moreover, the absence of a permanent magnet would result in the response being an octave above the transmitted signal.

The resistance of the transmitter is low, of the order of an ohm, whereas that of the receiver is of the order of 100 ohms, since the coils of the latter are wound with many turns of fine wire. The transformer passes the fluctuations from the line to the receiver while, at the same time, preventing the high resistance of the receiver from reducing the current in the line.

Mutual inductance

We shall now explain how mutual induction, illustrated in Fig. 4.8, can be represented quantitatively.

If an induced e.m.f. of 1 volt is induced in the secondary by a current in the primary charging at the rate of 1 ampere per second, the circuits are said to possess a mutual inductance of 1 henry.

The mutual inductance, M (in henrys), may be defined by the equation

$$M = \frac{-E}{dI/dt},$$

where E = e.m.f. induced in the secondary (in volts), and dI/dt = rate of change of the current in the primary (in amperes per second). When primary and secondary are interchanged M is the same.

The negative sign makes M a positive quantity because E is always in opposition to dI/dt, and hence of opposite sign.

If the current in the primary changes by 1 ampere in 1 second then, in the above equation for M, $dI/dt = 1$ and $M = -E$. But E is equal to the change in flux linkage through the secondary in 1 second. Thus

M = *magnetic flux linkage through the secondary, due to a primary current of 1 ampere.*

This may be regarded as an alternative definition of M and is useful in calculating mutual inductances. That the units are equivalent can be shown as follows:

$$\text{Henry} = \frac{V}{A\,s^{-1}} = \frac{V\,s}{A} = \frac{Wb}{A}.$$

Mutual inductance of two coaxial solenoids with an air core

Suppose a short secondary coil is wound round the outside and in the middle of a long primary coil (Fig. 4.23). It is possible to calculate the mutual inductance from the dimensions of the coils.

Let I = current through the primary, n = number of turns *per metre*

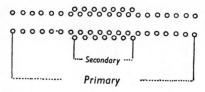

Fig. 4.23

of the primary, A = area of cross-section of the primary, N = *total* number of turns of the secondary.

Magnetic flux density inside the primary = $\mu_0 nI$.

Magnetic flux through the primary = $\mu_0 nIA$.

$\therefore$ Magnetic flux linkage through the secondary = $N\mu_0 nIA$.

Mutual inductance, M = magnetic flux linkage when $I = 1$ ampere.

$\therefore M = \mu_0 nNA$ henrys.

It should be noted that the primary coil has been assumed to be long; if it is not long the formula becomes more complicated.

<small>EXAMPLE</small>

A straight solenoid, 50 cm long, is wound with 1000 turns of mean diameter 4 cm. A secondary coil of 1200 turns is wound round the middle of the solenoid. Calculate the mutual inductance.

$$M = \mu_0 nNA,$$

$$\mu_0 = 4\pi \times 10^{-7} \text{ H m}^{-1}, \qquad n = \frac{1000}{50 \times 10^{-2}} \text{ m}^{-1},$$

$$N = 1200 \text{ turns}, \qquad A = \pi \times 2^2 \times 10^{-4} \text{ m}^2$$

$$\therefore M = 4\pi \times 10^{-7} \times \frac{1000}{0\cdot5} \times 1200 \times 4\pi \times 10^{-4}$$

$$= 0\cdot0038 \text{ H or } 3\cdot8 \text{ mH}.$$

Calibration of a ballistic galvanometer by means of a standard mutual inductance

A known mutual inductance can be used to calibrate a ballistic galvanometer (Fig. 4.24). The ballistic galvanometer is connected to the secondary of the mutual inductance and its throw is noted when a known current I ampere, measured by an ammeter, is reversed in the primary. The resistances of the ballistic galvanometer and of the secondary must be known; suppose that the total resistance of the secondary circuit is R. The quantity of electricity Q induced in the secondary

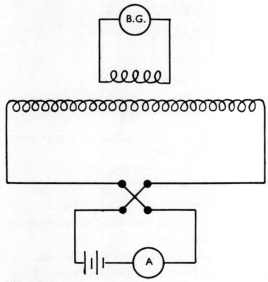

Fig. 4.24

when the current in the primary is reversed, is double that when the current in the primary is switched off.

$$Q = -\frac{2N\Phi}{R} \quad \text{(p. 81)},$$

where $N\Phi$ is the magnetic flux linkage with the secondary due to the current in the primary.

If M is the mutual inductance (equal to $\mu_0 nNA$ as explained above)

$$N\Phi = MI$$

$$\therefore Q = -\frac{2MI}{R}.$$

Knowing M, I and R it is possible to calculate Q. If θ divisions is the throw of the galvanometer, the sensitivity is θ/Q divisions per coulomb.

Avoidance of damping correction

We explained on p. 83 that the damping of a ballistic galvanometer increases as the resistance of the galvanometer circuit is reduced. The necessity for correcting for damping can be avoided by keeping the resistance of the galvanometer circuit constant in both the measuring and calibrating parts of an experiment. Thus, if the galvanometer is to be used to measure the quantity of electricity induced in a coil, the coil and the secondary of the mutual inductance are put in series and kept in the circuit during both parts of the experiment; the two damping corrections are then the same and cancel out.

Damping can be reduced by using a special double switch which breaks the galvanometer circuit immediately after breaking the primary circuit and after the discharge has passed through the galvanometer coil, thereby preventing eddy currents from flowing in the galvanometer coil.

Self-inductance

We explained on p. 88 that a changing current induces an e.m.f. in the coil or circuit in which it is flowing.

If an e.m.f. of 1 volt is induced when the current changes at the rate of 1 ampere per second, the coil or circuit is said to have a self-inductance of 1 henry.

The self-inductance, L (in henrys), may be defined by the equation

$$L = \frac{-E}{dI/dt},$$

where E = induced e.m.f. (in volts), and dI/dt = rate of change of current (in amperes per second). By reasoning similar to that used in the case of mutual inductance it can be shown that

L = magnetic flux linkage due to a current of 1 ampere.

Inductance of an air-cored solenoid

The self-inductance, often abbreviated to inductance, of an air-cored solenoid can be calculated from its dimensions. We will consider a long solenoid and neglect the effect of its ends, i.e. assume that the field inside it is the same as that of an infinitely long solenoid.

Let N be the total number of turns of the solenoid, l its length, A its area of cross-section, and I the current flowing.

Magnetic flux threading each turn of the solenoid = $\mu_0 \dfrac{NIA}{l}$.

Magnetic linkage in the solenoid = $\mu_0 \dfrac{NIA}{l} N$.

L = Magnetic flux linkage when I = 1 ampere

$= \dfrac{\mu_0 N^2 A}{l}$.

The expression applies accurately to a toroid of mean circumference l.

If we consider a single turn solenoid, consisting of a sheet of metal, for which $N = 1$, $A = 1$ and $l = 1$, then $L = \mu_0$. Thus μ_0 is the inductance in henrys of a 'unit solenoid'; its units are henrys per metre.

EXAMPLE

Calculate the self-inductance of a straight solenoid 50 cm long, wound with 1000 turns of mean diameter 4 cm, neglecting the effects of the ends of the solenoid.

$$L = \frac{\mu_0 N^2 A}{l}$$

$$= \frac{4\pi \times 10^{-7}.\, 1000^2.\, \pi \times 2^2 \times 10^{-4}}{0.5}$$

$$= 0.0032 \text{ H or } 3.2 \text{ mH.}$$

Summary

In the space surrounding an electric current there is a magnetic field which can be represented by lines of magnetic flux. Magnetic flux is measured in terms of the e.m.f. it induces when it changes in the neighbourhood of a circuit or a conductor.

Induced e.m.f. = rate of change of magnetic flux linkage.

$$E = -N\frac{d\Phi}{dt}.$$

Change of magnetic flux linkage = induced e.m.f. × time.

$$N\Phi = -\int E\, dt.$$

Φ is measured in webers or volt-seconds.

Magnetic flux density, $B = \dfrac{\Phi}{A}$.

Quantity of electricity induced (in coulombs) $= -\dfrac{N\Phi}{R}$.

Laws of electromagnetic induction:

(1) The direction of the induced e.m.f. is such that it tends to oppose the change to which it is due.

(2) The magnitude of the induced e.m.f. is equal to the rate of change of the magnetic flux linkage:

$$E = -N\frac{d\Phi}{dt}.$$

The alternating e.m.f. induced in a coil rotating uniformly in a uniform magnetic field is given by

$$E = E_0 \sin 2\pi ft,$$

where $\quad E_0 = BAN\, 2\pi f$

The units of e.m.f. are volts.

Mutual inductance, $M = -\dfrac{E}{dI/dt}$.

For coaxial solenoids, $M = \mu nNA$.

Self-inductance, $L = -\dfrac{E}{dI/dt}$.

For a solenoid, $L = \dfrac{\mu N^2 A}{l}$.

The unit of inductance is the henry.

Questions

1. Explain what is meant by a line of magnetic flux.
The direction of the current in a vertical solenoid is clockwise as seen from above. What is the positive direction of the magnetic flux (a) inside the solenoid, (b) outside the solenoid?

2. State the laws of electromagnetic induction.
The current in a vertical solenoid is clockwise as seen from above. A closed secondary coil is wrapped round the middle of the solenoid. State and explain the direction of the induced current in the secondary coil, as seen from above, when the current in the solenoid is (a) switched off, (b) switched on.

3. Define the weber.
Calculate the e.m.f. induced in a coil of 100 turns when the magnetic flux threading through it changes at the rate of 25 mWb s^{-1}.

4. Prove that the quantity of electricity induced in a circuit by a change of magnetic flux is independent of the time taken for the change to occur.
A metal-framed casement window, of length 120 cm and breadth 50 cm, lies in the eastern wall of a house. Regarding the window frame as a conducting circuit of resistance 1·0 mΩ, calculate the quantity of electricity which will pass through the frame as it swings open through 90°. The horizontal component of the earth's magnetic flux density is $1\cdot8 \times 10^{-5}$ Wb m^{-2}.

5. Describe how the magnetic flux density between the poles of an electromagnet may be measured by means of a search coil and a ballistic galvanometer.
A search coil, of 200 turns and area of cross-section 4 cm², is perpendicular to the field between the poles of an electromagnet and is then jerked away. The coil is connected to a ballistic galvanometer, which registers a throw of 40 μC. If the

total resistance of the coil and ballistic galvanometer circuit is 500 Ω, find the flux density of the magnetic field of the electromagnet.

If, instead of being jerked out of the field, the search coil were turned through 180°, what charge would flow through the galvanometer?

6. Two galvanometers, A and B, available for use with a search coil of resistance 5 Ω, have the following constants:

	Resistance (ohms)	Sensitivity (divisions per microcoulomb)
A	10	70
B	500	350

Which of these will be the more sensitive when used with the coil and to what extent?
(N)

7. An aeroplane of wing-span 30 m flies horizontally at a speed of 1000 km/hour. What is the p.d. between the wing tips if the vertical component of the earth's magnetic field is 4.0×10^{-5} Wb m^{-2}?

8. A railway train is travelling at 60 miles per hour along a level line. The vertical component of the earth's magnetic field is 4.2×10^{-5} Wb m^{-2}. Find the e.m.f. induced in each axle, if the length of an axle is 4 ft 9 in. (1 ft = 30·5 cm.)

. 9. Find the difference of potential between the two ends of a horizontal wire 1 m long pointing E and W, 2 s after it has been dropped from rest. The horizontal component of the earth's magnetic field is 2.0×10^{-5} Wb m^{-2}.

10. An electric fan, of radius 10 cm, rotates at 800 revolutions per minute with its axis horizontal and parallel to the magnetic meridian. What is the p.d. between the axis and the rim if the horizontal component of the earth's magnetic field is 1.8×10^{-5} Wb m^{-2}?

11. A solid metal wheel 1 m in diameter spins in a vertical plane at right angles to the magnetic meridian, and makes 300 revolutions per minute. It is found that there is a p.d. between its centre and its rim of 71 μV. What is the flux density of the earth's horizontal magnetic field?

12. The magnetic flux between the two poles of a dynamo is 25 mWb. At what rate must the armature, consisting of 100 turns of wire, be revolved for the dynamo to generate a mean e.m.f. of 100 V? (*Hint.* The magnetic flux threads the coil in opposite directions each half revolution; hence the coil can be regarded as threaded twice in each half revolution.)

13. The flux density between the pole pieces of a powerful electromagnet is 2 Wb m^{-2}. What is the force exerted on 1 cm of wire carrying a current of 5 A when the wire is (*a*) at right angles to the field, (*b*) parallel to the field, (*c*) at an angle of 30° to the field?

14. A horizontal cable running E and W carries a current of 1000 A. The earth's magnetic field, running N and S but dipping at an angle of 67° with the horizontal, has a magnetic flux density of 46×10^{-6} Wb m^{-2}. What is the magnitude of the force per metre on the cable and its direction?

15. A straight wire, of length 20 cm and resistance 0·25 Ω, lies at right angles to a magnetic field of flux-density 0·40 Wb m⁻², and is free to move. A constant p.d. of 2·0 V is applied to its ends.

Calculate the force on the wire when it is at rest and when it is moving in the direction of the force at a speed of 15 ms⁻¹. What is the maximum speed of the wire?

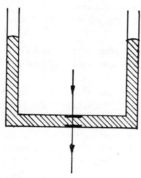

Fig. 4.25

16. Mercury is contained in a U-tube of uniform square cross-section. Electrodes are sealed inside the upper and lower walls of the horizontal arm of the U-tube (Fig. 4.25). What will happen when a current is passed between the electrodes in the direction shown, and a magnetic field is applied across the bottom of the U-tube, perpendicular to, and down into, the paper in Fig. 4.25?

If the side of the square cross-section of the U-tube is 4 mm, the current is 10 A and the magnetic flux density of the field is 0·5 Wb m⁻², calculate the difference in the mercury levels in the vertical arms of the U-tube. The density of mercury is 13 600 kg m⁻³.

17. Calculate the torque on a rectangular coil of 50 turns, of dimensions 2·0 cm × 3·0 cm, carrying a current of 15 mA in a magnetic field of flux density 0·60 Wb m⁻², when its plane is (a) parallel to the field, (b) at 30° to the field.

18. The coil of a galvanometer has 300 turns of mean area 1 cm². A couple of 10⁻⁷ N m applied to the coil causes it to twist through 1 revolution against the torsional control of the suspension. If there is a radial field of 0·2 Wb m⁻², what current, when passed through the galvanometer, will cause a spot of light on a scale 1 m away to be deflected through 1 mm?

19. Describe the construction of a radial-field, moving-coil galvanometer and discuss the factors which affect its sensitivity. Why does the coil of such a galvanometer rapidly cease to oscillate when its terminals are short-circuited?

20. Copper coils carrying excessively large currents have been known to break up. Explain why this occurs.

21. State the laws of electromagnetic induction and describe how you would verify them by experiments.

22. Give a brief explanation of each of the following and state the important general principle illustrated by them:

(*a*) The oscillations of a moving-coil ballistic galvanometer can be stopped very quickly by short-circuiting the coil.

(*b*) When a plate of copper or aluminium is pushed into a strong magnetic field between the poles of a powerful electromagnet, considerable resistance to the motion is felt, but no such effect is observed with a sheet of glass.

(*c*) A thick copper ring can be made to jump off the pole of an electromagnet if a large a.c. is used. (CS)

23. Describe the construction of a transformer and of an induction coil, and explain the essential differences between them. What is the function of the capacitor in the induction coil? (OS)

24. The primary of a transformer is connected through a fuse to a source of supply, the secondary being on open circuit. If the source is d.c. the fuse is blown, but not if the source is a.c. Why is this? (C)

25. What ratio of turns is required in a transformer to step 230 V down to 20 V? If the secondary current is 1·0 A, what is the primary current, assuming 100% efficiency? What is the power output of the transformer?

26. An alternating current transformer has a primary of N_1 turns connected to the a.c. supply, and a secondary of N_2 turns. Discuss in general terms the relation between (*a*) the primary and secondary voltages when the secondary circuit is open, (*b*) the primary and secondary currents when the secondary circuit is closed. Why does the closing of the *secondary* circuit cause an increase in the *primary* current? (O & C)

27. Describe the construction and mode of action of an a.c. transformer.

A power station has an output of 5000 kW at 400 V. The power is transformed up to 120 000 V for transmission along a line of total resistance 25 Ω, and is then transformed down to 200 V. If each transformer has an efficiency of 96%, what power output and current will be available in the 200 V circuit? (O & C)

28. Describe a simple a.c. generator and explain the principles on which its action depends.

What are the chief sources of energy loss in a practical a.c. generator and how are they reduced to a minimum? (C)

29. Describe and explain the working of a simple permanent magnet dynamo such as is used for bicycle lighting. How does the current through the lamp change as the speed increases? (OS)

30. Explain the construction and action of a simple form of series-wound d.c. dynamo.

Account for the fact that the machine will run as a motor if supplied with d.c. of suitable voltage.

Explain what would happen if you connected the machine to a source of a.c. of the same voltage. (O)

31. The resistance of a certain d.c. motor is 0·25 Ω. When running with no load on a 100 V supply it takes 1·5 A, whilst on full load the current rises to 35 A. What is the back e.m.f. in each case and the electrical efficiency in the second case? (D)

32. What do you understand by the back e.m.f. of an electric motor?

A shunt-wound d.c. electric motor takes a current of 100 A from 200 V mains. The shunt-field coils have a resistance of 40 Ω; the armature has a resistance of 0·5 Ω. Find (a) the back e.m.f. and hence (b) the electrical energy converted into mechanical work per second. Verify that (b) is the difference between the total power supplied and that turned into heat. (C)

33. A shunt-wound d.c. motor has a variable resistance connected in series with its field winding, while the armature is connected directly to the supply mains. It is found that, as the resistance in the field circuit is *increased*, the speed of the motor also *increases*. Explain this phenomenon and suggest a reason why this might not be a good method of controlling the speed of the motor over a wide range of values. (CS)

34. A coil of 300 turns, each of area 100 cm², is rotated at 600 revolutions per minute about an axis at right angles to a magnetic field of flux density 0·20 Wb m^{-2}. Calculate the e.m.f. induced in the coil when its plane is at an angle of (a) 0°, (b) 30°, (c) 60°, (d) 90° with the field.

35. A coil of 50 turns and of radius 10 cm makes 1200 revolutions per minute in a magnetic field of flux density 0·020 Wb m^{-2}. Calculate the average value of the e.m.f. in the coil in volts (irrespective of sign). (C adapted.)

36. A circular coil of 100 turns, each of radius 10 cm, is rotated 10 times per second about an axis at right angles to a magnetic field of 0·10 Wb m^{-2}. Find the position of the coil when the e.m.f. across its ends is a maximum, and the value of this e.m.f.

If, when in this position, a current of 1 A is taken from the coil, what is the torque required to maintain its angular velocity? (O & C adapted.)

37. A small flat circular coil of radius r and of n turns is situated at the centre of a larger flat coil of radius R and N turns carrying a current of I. The small coil is rotated at an angular velocity ω about an axis in its own plane and in that of the larger coil. Show that there is induced across the end of the small coil a single-phase alternating e.m.f., and find its peak value in volts, and its frequency. (Take $\mu_0 = 4\pi \times 10^{-7}$ H m^{-1}.)

38. Explain what is meant by the statement that two circuits have a mutual inductance of 1 henry.

The mutual inductance of the two coils of an induction coil is 30 H. A current of 1·5 A in the primary is extinguished in 0·002 s. Calculate the e.m.f. induced in the secondary coil.

39. Explain what is meant by the statement that a circuit has an inductance of 1 H.

A current of 2 A through a given coil provides 1·0 Wb turns of flux linkage. What is the inductance of the coil?

40. (a) An air-cored coil has a self-inductance of L. If it is tapped at its centre, what is the self-inductance of each half of the coil? Where would the tapping have to be to produce an inductance of $\frac{1}{2}L$?

(b) Explain why, if a wire carrying a high-frequency a.c. is wrapped round a pencil, the current is reduced.

How could the wire be wrapped without reducing the current? (*Contd. over.*)

(c) A coil in the form of a spiral spring is stretched to double its length. What effect does this have upon its inductance?

41. Calculate the self-inductance of a solenoid 50 cm long and 4 cm diameter, wound with 5000 turns of wire, neglecting the effects of the ends.

42. Explain why iron-cored solenoids are employed for high inductances. Why does the inductance tend to decrease in value as the current in the solenoid is increased?

43. Explain why the mutual inductance of two coils, having their axes in the same straight line, becomes greater as the two coils approach closer together. Use this fact to account for the source of the energy when the two coils are placed in series with a battery and they approach as a result of the force of attraction between them.

44. A solenoid 50·0 cm long, and of mean diameter 3·00 cm, is uniformly wound with 1000 turns. A secondary coil of 800 turns is wound closely round the middle of the solenoid and is connected to a ballistic galvanometer. The total resistance of the secondary circuit is 420 Ω. When a current of 3·00 A is reversed in the solenoid the throw of the galvanometer is 12·0 divisions. Calculate the sensitivity of the galvanometer in divisions per microcoulomb, assuming the damping of the galvanometer to be negligible.

45. Explain in simple terms the following observations:

(i) Iron filings placed in a magnetic field tend to set themselves in the direction of the field.

(ii) When a soft iron core is placed inside a solenoid connected to a 12 V d.c. supply, the magnetic field outside the solenoid is increased considerably but the current through the solenoid is unchanged.

(iii) If, however, the supply is 12 V, 50 Hz a.c., the insertion of the core reduces the value of the circulating current.

An air-cored solenoid, 0·8 m long and of cross-sectional area 10^{-3} m^2, has 1600 turns. Calculate (a) the flux linked with the solenoid when the current through the solenoid is 5 A, and (b) the induced back e.m.f. at the instant when the current is increasing at the rate of 100 A s^{-1}.

(Take the value of μ_0 to be $4\pi \times 10^{-7}$ H m^{-1}.)   (O 1971)

46. Give an account of the moving coil ballistic galvanometer. Explain how this instrument is used for the measurement of quantities of magnetic flux.

For calibration purposes, a ballistic galvanometer is connected to the secondary of a mutual inductance of $1·2 \times 10^{-5}$ H, the total resistance of the secondary circuit being 10^3 Ω. When a current of 2·0 A is reversed in the primary, the meter deflects through 30 divisions. What is the charge sensitivity in coulomb per division?

A charged 25 μF capacitor is connected to the ballistic galvanometer through a 4000 Ω resistance for 0·1 s (which is much shorter than the period of swing of the ballistic galvanometer) and again the deflection is 30 divisions. What was the initial charge on the capacitor?   (O 1970)

47. State the laws of electromagnetic induction and describe briefly experiments to show their validity.

A coil A passes a current of 1·25 A when a steady potential difference of 5 V is maintained across it, and an r.m.s. current of 1 A when it has across it a sinusoidal potential difference of 5 V r.m.s. at a frequency of 50 Hz (cycles per second). Explain why the current is less in the second case, and calculate the resistance and the inductance of the coil.

The same coil A, which has 100 turns, has a second coil B with 500 turns wound on it so that all the magnetic flux produced by A is linked by B. Find the r.m.s. value of the e.m.f. that appears across the open-circuit ends of B when a sinusoidal alternating current of 1 A r.m.s. at a frequency of 50 Hz is passed through A. Why is the ratio of this e.m.f. to the r.m.s. potential difference across A not the same as the ratio of the number of turns in B and A, i.e. 5:1?

Explain why the insertion of an iron core into the coils would decrease the current in A and increase the e.m.f. across B, if the alternating potential difference across A were kept unchanged; the effects of hysteresis and eddy currents in the iron may be neglected.   (O & C 1968)

48. Explain the terms *potential, potential gradient, potential difference, electromotive force*. In current electricity it is necessary to distinguish carefully between p.d. and e.m.f.; why is this not the case in electrostatics?

Explain how the potentiometer principle may be applied

(*a*) to compare the e.m.f.'s of two d.c. sources or generators;

(*b*) to compare the values of two low resistances.

Describe briefly the Lorenz method for the absolute determination of resistance.
                                                                        (O 1968)

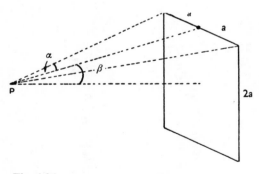

Fig. 4.26

49. Fig. 4.26 shows a long uniformly wound solenoid P round the middle of which a short secondary coil S is closely wound. The dimensions and other relevant details are:

For P: length 50 cm (0·5 m)
           area 30 cm² (3×10^{-3} m²)
           number of turns 1500
           resistance 5 Ω.

For S: number of turns 500
           resistance 300 Ω
           resistance of attached external circuit 5000 Ω.

Giving in each case an explanation of your method, calculate:

(a) the self-inductance of P;

(b) the mutual inductance between P and S;

(c) the charge that circulates round S and its circuit when a current of 1 A is reversed in P;

(d) the current flowing in P when it is removed from S and connected to a 12 V, 50 Hz alternating-current supply.

(Take the value of μ_0 to be $4\pi \times 10^{-7}$ H m^{-1}.)   (O 1968)

5   Magnetic properties of matter

If an iron core is inserted in a solenoid carrying a current, the magnetic flux inside the solenoid is greatly increased. This can be demonstrated by wrapping turns of wire round the middle of the solenoid to form a secondary coil, connecting the secondary coil to a ballistic galvanometer and switching off, or reversing, the current in the solenoid. The quantity of electricity induced in the secondary coil is proportional to the change in the magnetic flux threading the secondary coil. With air inside the solenoid there is a small kick in the ballistic galvanometer but, with an iron core, a very much greater kick.

> Let B_0 = magnetic flux density in a long solenoid with air core (or vacuum).
>
> B = magnetic flux density with iron core.
>
> J = magnetic flux density provided by the iron.
>
> $B = B_0 + J$.

J is called the *magnetic polarization* of the iron and it is defined as the magnetic flux density provided by the iron. Like B and B_0 it is measured in webers per square metre and it is a vector quantity.

Magnetic flux is provided by iron because its atoms are equivalent to tiny loop currents, due to revolving or spinning electrons. Groups of many millions of atoms, known as domains, act as individual powerful loop currents. When the iron is unmagnetized the magnetic axes of the domains are oriented at random and hence their magnetic effects neutralize each other. But when the iron is placed in an external magnetic field some of the domains become aligned with their magnetic axes in the direction of the field and thereby provide, as it were, internal ampere-turns which may be far greater than the external ampere-turns.

Permeability

The ratio B/B_0 is called the *relative permeability*, μ_r, of the iron:

$$\frac{B}{B_0} = \mu_r.$$

Here B and B_0 are the flux densities in a long solenoid when it is completely filled with air and with iron (or other magnetic material) respectively. μ_r is a pure ratio and has no units.

Materials which have a high μ_r are known as ferromagnetic; they comprise iron, nickel, cobalt, gadolinium and certain alloys. The value of μ_r is not constant but varies with B. For cast iron, which is not a very good magnetic material, μ_r has a maximum value of about 350. A silicon steel, stalloy, which is widely used in a.c. generators and transformers, has a maximum μ_r of about 6000. Some nickel–iron alloys have values of μ_r up to 100 000, but they require careful heat treatment and are susceptible to mechanical strain.

The magnetization curve

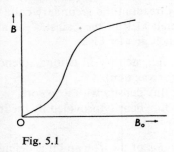

Fig. 5.1

The B/B_0 graph for a ferromagnetic material has a characteristic shape (Fig. 5.1). The values of B and B_0 can be obtained by the experiment indicated in Fig. 5.2.

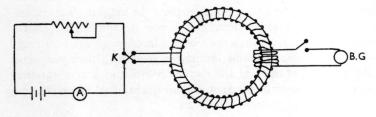

Fig. 5.2

The material, say iron, is made in the form of a ring and is wrapped completely and closely round with a ring solenoid, called a toroidal winding. The advantage of a ring is that the solenoid has no ends, and the magnetic flux is confined to the ring and does not emerge into the air. A ready-wound straight solenoid and a straight bar of the material are more commonly used, because of their greater convenience, but, as the corrections to be applied are somewhat complicated, we shall confine our discussion to the simpler case of a ring.

The ring solenoid is connected through a reversing switch, K, to a battery, rheostat and ammeter. If r is the mean radius of the ring, so that $2\pi r$ is its mean circumference, and N is the total number of turns, the magnetic flux density, B_0, inside the solenoid for a current I is $\mu_0 NI/2\pi r$. The radial thickness of the ring must be small compared with r (not greater than one-seventh); otherwise the solenoid is appreciably more densely wound on its inner circumference than on its outer, and an appreciable error is introduced by using the mean circumference, $2\pi r$, in the calculation of B_0.

B is measured by means of a secondary coil wound round part of the ring and connected to a ballistic galvanometer (a resistance being inserted in series if the throw of the galvanometer is too large).

The iron ring is first demagnetized by passing the maximum current through the solenoid and continually reversing the current while it is gradually reduced to zero. The explanation of this procedure for demagnetization is given on p. 117.

The rheostat is adjusted so that the smallest suitable current flows through the solenoid and the current is reversed about 20 times to put the iron in a cyclic condition. The necessity for this is also explained on p. 117. The switch in the secondary circuit is now closed and the current reversed once more; the throw in the galvanometer is proportional to $2B$, where B is the magnetic flux density in the ring.

The procedure is repeated for several values of the current until the maximum current is reached. If A is the area of cross-section of the ring, the total flux is BA. The quantity of electricity induced in the secondary circuit is $2(N'BA/R')$ where N' is the number of turns in the secondary coil and R' is the total resistance of the secondary circuit (p. 81). If the ballistic galvanometer has been calibrated (for the value of R') the quantity of electricity corresponding to a given throw is known and hence B can be calculated. If the ballistic galvanometer has not been calibrated, a curve of throw against B_0 can be plotted and this will have a shape similar to that of the B/B_0 curve. The resistance, R', of the secondary circuit should not then be changed during the experiment.

When the iron is completely magnetized so that it can contribute no further magnetic flux, it is said to be *saturated*. This state is represented by the top of the curve, which is nearly horizontal.

Variation of permeability

If the ring in Fig. 5.2 had a non-magnetic core, such as wood, the B/B_0 curve would be a straight line (of very small slope), showing that the permeability, B/B_0, of wood is constant. The curve for a ferromagnetic material (Fig. 5.1) has a variable slope and hence μ_r varies. The way μ_r for iron varies with B_0 is shown in Fig. 5.3.

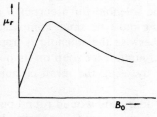

Fig. 5.3

Hysteresis

If the magnetizing current is gradually varied through a complete cycle from the positive maximum through zero to an equal negative maximum and back again to its positive value, the B/B_0 curve obtained is shown in Fig. 5.4 (which includes also the magnetization curve).

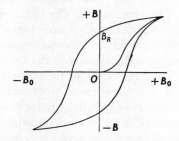

Fig. 5.4

The complete outer curve is known as the hysteresis loop or cycle. The word 'hysteresis' means 'to lag behind', and it will be seen that B lags behind B_0, becoming zero after B_0 has done so. The study of hysteresis has great practical importance because the iron in a transformer or a.c. generator or motor is subjected to a continuously alternating magnetizing force.

In Fig. 5.4, B_R represents the magnetic flux retained by the material when B_0 has been reduced to zero from a positive maximum value; it is called the *remanent flux density*. It is greater the greater the maximum value of B_0, and its limiting value is known as the *remanence* or *retentivity* of the specimen.

It is clear that the magnetic condition of a material depends on its previous magnetic history. Thus, in Fig. 5.4, for every positive value of B_0 there are three possible values of B in the specimen. It is therefore essential to demagnetize a specimen before experimenting upon it, and this can best be done by repeated reversals of B_0, while B_0 is steadily reduced. By this means the specimen is taken through diminishing cycles, as in Fig. 5.5.

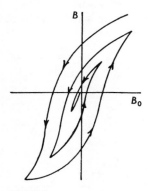

Fig. 5.5

If, after demagnetization, the specimen is taken round the hysteresis cycle several times, using the same maximum value of B_0 in each case, the maximum value of B increases. In other words, the loops do not close. Only after the specimen has been taken round the hysteresis cycle 20 or 30 times will the loops close; the specimen is then said to be in a cyclic condition.

Experimental determination of the hysteresis loop

To obtain the hysteresis loop for iron, say, in the form of a ring, the apparatus in Fig. 5.6 may be used. It will be seen that the circuit differs from Fig. 5.2 only by the inclusion of a switch S which, when closed, short-circuits the variable resistance R. As before, a resistance must be placed in series with the ballistic galvanometer if its sensitivity is too high.

The iron is first demagnetized by continual reversal of the current

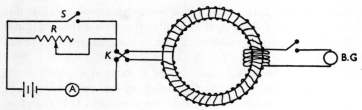

Fig. 5.6

in the solenoid, by means of K, as the current is reduced to zero (S being open and the value of R being gradually increased). S is closed and the battery should have an e.m.f. sufficient to send a current through the solenoid which will produce magnetic saturation of the iron, corresponding to point 1 in Fig. 5.7. The current is reversed about 20 times to bring the iron to a cyclic condition.

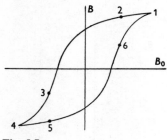

Fig. 5.7

The switch in the secondary circuit is now closed and the current reversed once more. The throw of the ballistic galvanometer will be proportional to $2B_1$, or $B_1 + B_4$, where B_1 and B_4 represent the magnetic flux densities at the points 1 and 4. Suppose we are now at point 1 (rather than point 4). The switch S is opened, thereby bringing into operation the resistance R, and this causes the current in the solenoid to be reduced to that corresponding to point 2. The throw of the galvanometer is proportional to $B_1 - B_2$.

Keeping R constant, six points can be obtained as follows:

(a) Open S, go from 1 to 2, throw $\propto B_1 - B_2$ (as already explained).

(b) Reverse K, go from 2 to 3, throw $\propto B_2 + B_3$.

(c) Close S, go from 3 to 4, throw $\propto B_4 - B_3$.

(d) Open S, go from 4 to 5, throw $\propto B_4 - B_5$.

(e) Reverse K, go from 5 to 6, throw $\propto B_5 + B_6$.

(f) Close S, go from 6 to 1, throw $\propto B_1 - B_6$.

It is essential that the switching operations should be performed in the proper order, so that we proceed in sequence round the hysteresis loop; otherwise the iron will be taken round a minor asymmetrical loop and the cyclic process broken. Thus, if a switching operation is made in the wrong order, the experiment must be started again.

Other sets of points can be obtained by altering for each set the resistance R, and in this way, sufficient points obtained to plot the whole loop.

Hysteresis loss

The shape and size of the hysteresis loop depends on the nature of the material of the specimen; the loop is narrow for soft iron and wider for steel (Fig. 5.8). The area of the loop represents the energy expended in the material in going through the cycle; this energy appears in the form of heat in the specimen.

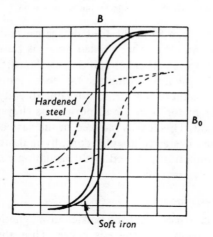

Fig. 5.8

The shaded areas in Fig. 5.9 represent the work done as the material is taken round the complete hysteresis cycle. In Fig. 5.9 (a) the current is reduced from a positive maximum to zero and, since the induced e.m.f. tends to prevent the current from decreasing, it acts in the same direction as that of the source of supply; thus energy is returned to the electrical circuit. In Fig. 5.9 (b) the current is increased from zero to a negative maximum. The induced e.m.f. tends to stop the current from growing, it is therefore in opposition to the source of supply and energy must be provided by the source of supply. In Fig. 5.9 (c) the current is once more reduced to zero and energy is returned to the electrical circuit. In Fig. 5.9 (d) the current increases to a positive maximum, thereby completing the hysteresis cycle, and energy is taken from the

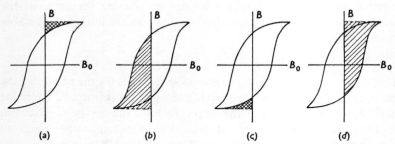

(a)   (b)   (c)   (d)

Fig. 5.9 (a) Energy returned to circuit. (b) Energy taken from circuit.
(c) Energy returned to circuit. (d) Energy taken from circuit.

electrical circuit. It will be seen from the four diagrams, taking cross-hatched areas as negative and the other shaded areas as positive, that the net energy taken from the electrical circuit during the whole hysteresis cycle is represented by the area within the hysteresis loop.

Besides the hysteresis loss there is a further loss due to eddy currents which are induced in the material in the same way as they would be induced in a secondary coil. This loss is reduced in the case of a transformer for example, which has a rectangular core, by making the core of thin sheets, called laminations, coated with insulating material to reduce the flow of the eddy currents.

Use of hysteresis loop

The suitability of a specimen of iron for a particular magnetic purpose can be seen at a glance from its hysteresis curve. Thus the steel in Fig. 5.8 is more suitable than the soft iron for a permanent magnet; the soft iron, on the other hand, is more suitable for the core of a transformer, because its hysteresis loss is low and it has a higher flux density for a given value of B_0.

The magnetic properties of iron can be varied considerably by alloying it with other metals. Iron alloyed with $3\frac{1}{2}\%$ silicon is known as Stalloy and is widely used for transformer cores. It has a maximum permeability of about 6000, a resistivity five times that of iron, with the advantage of reducing eddy current losses, and a hysteresis loss of about 300 J m^{-3} per cycle.

Alloys of nickel and iron are called permalloys, but the term is usually applied to the particular alloy containing 78% Ni. This has the large maximum relative permeability of about 100 000 and the low hysteresis loss of 20 J m^{-3} per cycle. Its relative permeability can be increased very considerably beyond 100 000 by careful annealing, i.e. cooling slowly to prevent internal strains, but it is very sensitive to heat treatment and to magnetic and mechanical shocks. Its principal use is the 'loading' of submarine cables. It becomes highly saturated in the earth's magnetic field alone.

The only material available for permanent magnets at one time was carbon tool steel. This was replaced by tungsten steels and then by cobalt steels. In recent years extremely powerful 'alnico' magnets have revolutionized the design of instruments and machines incorporating permanent magnets, such as meters, loud-speakers and magnetos.

The addition of manganese makes steel intensely hard and almost non-magnetic. No-mag is a nickel–manganese cast iron used for cable boxes, meter cases, etc., where a strong non-magnetic material is required.

Permanent magnets

A permanent magnet in an instrument or machine usually takes the form of a ring with an air gap between the poles. Permanent magnets operate on that part of the hysteresis loop in the second quadrant (Fig. 5.10), and the most important criterion of the usefulness of a

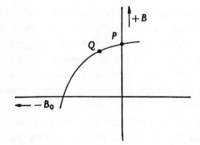

Fig. 5.10

material for a permanent magnet is the area under this part of the curve which represents the energy stored in the material.

Suppose that a ring of the material has been magnetized by a coil and that the coil is removed. So long as the material remains in a closed ring the remanent flux density is represented by the point P, but, when an air gap is cut, the point moves to Q, or lower if the air gap is made larger, because the internal 'ampere-turns', or magnetomotive force due to the atomic currents inside the magnet, must send the flux through the air gap, as well as through the iron, and hence the flux in the iron is reduced from that at P to that at Q.

If the magnet is jarred or subjected to some form of shock the point Q may move lower on the curve, and magnets are often deliberately demagnetized to a point lower than Q so that there is little chance of further demagnetization.

Diamagnetism and paramagnetism

All substances are found to have magnetic properties although these are usually very slight. In 1845 Faraday found that a rod of glass suspended by a fine silk fibre between the poles of a powerful electromagnet set itself at right angles to the magnetic field. The same happens to bismuth which exhibits the effect comparatively strongly, and to

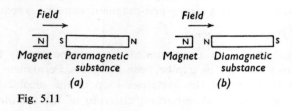

Fig. 5.11

many other substances including the tissues of the human body. As Faraday remarked, 'If a man could be in the magnetic field, like Mohammed's coffin, he would turn until across the magnetic line.'

Rods of platinum and aluminium set themselves parallel to the magnetic field. Faraday coined two new words to classify the two kinds of substances: *diamagnetics* set themselves across, and *paramagnetics* parallel to, the magnetic field.

The fundamental difference between the two is that paramagnetics become induced magnets in the direction of an external field, as iron does, whereas diamagnetics become induced magnets in the opposite direction (Fig. 5.11).

Paramagnetic substances are attracted by a magnet and diamagnetic substances are repelled, as we should expect from Fig. 5.11. A rod of a diamagnetic substance sets itself at right angles to the magnetic field between the poles of an electromagnet because its ends are repelled by the poles.

Explanation of diamagnetism and paramagnetism

Ampère suggested that magnetism might be due to circulatory electric currents in the molecules. These currents can now be identified as revolving or spinning electrons. Thus magnetism is a product of electricity; it cannot exist apart from electricity.

Diamagnetism was accounted for by Langevin in 1905 as follows. When a substance is placed between the poles of an electromagnet and the current is switched on, the change in magnetic flux through the molecular circuits of the substance will give rise to induced e.m.f.'s. These induced e.m.f.'s, by Lenz's law, will be in such a direction as to cause changes in the magnetic fields of the molecular currents which oppose the field of the electromagnet. The speeds of the revolving electrons can be regarded as being increased or decreased, according to the sense of their rotation, and since there is no resistance to their rotation, the effect will persist after the field of the electromagnet has become constant. When the current in the electromagnet is switched off or the substance is taken from between the poles, the molecular currents will return to their original values.

Thus all substances should exhibit diamagnetism; when placed in a magnetic field they should become magnetized in a direction opposite to that of the field.

In some substances, however, the effect is masked by another which gives rise to paramagnetism. Suppose the revolving electrons in the molecule do not counterbalance each other, and the molecule has a resultant magnetic moment. Each molecule will align itself in a magnetic field, like a magnetic compass, so that it increases the flux.

As would be expected from the theory, diamagnetism is independent of temperature but paramagnetism decreases with temperature, since more violent thermal agitation tends to impede alignment of the molecules.

Magnetic susceptibility

A useful concept for use with diamagnetic and paramagnetic materials is known as susceptibility. Suppose that the material is placed in a magnetic field of flux density B_0, that its magnetic polarization is J and that the resultant flux density due to the external field and to the material is B.

$$B = B_0 + J.$$

$$\therefore \frac{B}{B_0} = 1 + \frac{J}{B_0}.$$

J/B_0 is called the *magnetic susceptibility* of the material and is denoted by χ_m.

Since $\quad B/B_0 = \mu_r$ (p. 113)

$\mu_r = 1 + \chi_m$.

The magnetic susceptibility of diamagnetic materials is negative, because they become magnetized in a direction opposite to that of the external field, and μ_r is less than 1; in the case of paramagnetic materials χ_m is positive and μ_r is greater than 1.

The values of μ and χ_m for paramagnetics and diamagnetics differ from those for iron in that they are very much smaller and also constant, being independent of the field.

Some typical values are as follows:

Diamagnetics	χ_m	μ_r
Bismuth	−0·000 18	0·999 82
Silver	−0·000 019	0·999 981
Glass	−0·000 013	0·999 987
Water	−0·000 009	0·999 991
Paramagnetics		
Platinum	+0·000 25	1·000 25
Aluminium	+0·000 023	1·000 023
Saturated solution of $FeCl_3$ (at room temperature)	+0·000 64	1·000 64
Air (1 atmosphere)	+0·000 000 38	1·000 000 38

Ferromagnetism

The comparatively enormous relative permeabilities of ferromagnetic substances were first explained by Weiss in 1907. He suggested that a piece of iron, or of any ferromagnetic substance, is made up of small

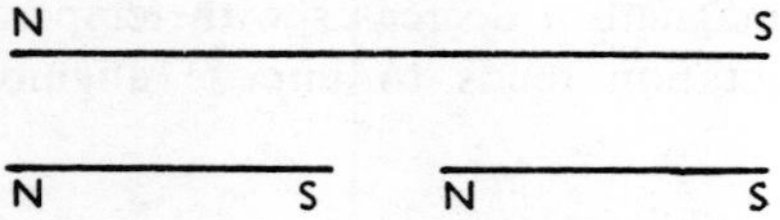

Fig. 5.12

regions, called domains, each of which is spontaneously magnetized to saturation because of the mutually aligning influence of the atoms. In unmagnetized iron the magnetic axes of the domains are oriented at random, so that there is no external magnetic effect. When the iron is magnetized by being placed in a magnetic field, the magnetic axes of some of the domains are aligned in the direction of the field.

This is similar in principle to the theories of Weber and Ewing in the latter part of the nineteenth century. It was known that, on breaking a bar magnet in two, a pair of fresh poles appears as in Fig. 5.12, and that this continues to happen if the bar magnet is broken into smaller and smaller pieces. Weber and Ewing therefore assumed that iron is composed of many tiny elementary magnets, but they assumed that these were the molecules of the iron. They explained the difference between unmagnetized and magnetized iron in the same way as Weiss: in unmagnetized iron the axes of the molecular magnets were oriented at random and in magnetized iron they were aligned.

The existence of domains has been confirmed by two pieces of experimental evidence. Barkhausen wrapped around an iron wire or strip a secondary coil connected to an amplifier and headphones (Fig. 5.13). As the magnetizing field (due to a solenoid not shown in the figure) is slowly increased, a series of clicks is heard in the headphones, each click being caused by a tiny current induced in the coil when the magnetic flux through it is increased by the alignment of the axis of a domain. The phenomenon is known as the Barkhausen effect.

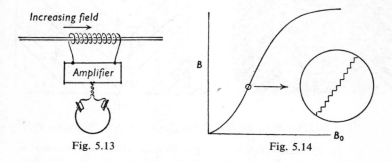

Fig. 5.13   Fig. 5.14

Thus, if the magnetization curve could be enormously magnified, it would be seen to be made up of a series of steps as in Fig. 5.14. From the sizes of the steps Bozorth was able to estimate the sizes of the domains. He found that the sizes were very variable, the average size being of side about 10^{-4} cm, so that each domain contains many millions of atoms.

The other piece of evidence for the existence of domains was discovered by Bitter. He sprinkled very fine iron filings on the polished surface of a ferromagnetic substance and found, when he examined the surface through a microscope, that the filings arranged themselves in patterns (Fig. 5.15). The domains behave like small permanent magnets and the filings settle where two of them come together.

The greatly increased knowledge of atomic structure since the time of Weiss has now made it clear that the atoms of iron, and of other ferromagnetic substances, happen to have critical dimensions and internal structure which enable electron spins in one atom to have a powerful aligning influence on neighbouring atoms. The forces between the atoms are called exchange forces. At a certain temperature, known as the *Curie point*, the thermal agitation of the atoms overcomes the exchange forces and the domains lose their magnetic properties. In the case of iron, the Curie point is about 1043 K; above this temperature iron is non-magnetic.

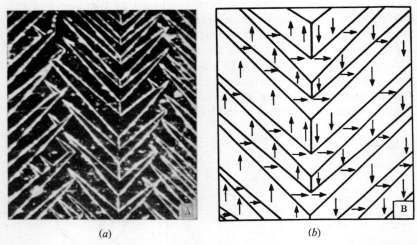

(a)   *(b)*

Fig. 5.15. (a) Bitter pattern on the surface of iron, magnified about 120×.

(b) The interpretation of the pattern in (a), the arrows showing the direction of domain magnetization. The surface was first made smooth by mechanical polishing with emery powder and rouge. If the surface were not smooth, the pattern would represent its relief rather than its magnetic state. Mechanical polishing, however, causes a state of strain in the surface; the strained surface was therefore removed by making it the anode in electrolysis.

The pattern is not very similar to Fig. 5.16, but the latter was a purely diagrammatic representation of the simplest orientations of the axial planes of the domains with respect to the surface.

A metal such as iron is composed of a large number of crystals, or crystal fragments, and there is usually a number of domains in each crystal fragment. When there is no external field, the magnetic axes of the domains in a crystal are aligned along one of six directions, positive and negative directions along three mutually perpendicular crystal axes.

The random orientation of the axes of the domains in an unmagnetized bar of iron is indicated in Fig. 5.16 (a). The domains are not regular in shape, however, nor are they equal in size. When the iron is magnetized by a weak magnetic field, domains whose magnetic axes are in the direction of the field become enlarged at the expense of domains with a less favourable orientation. This causes only weak magnetization of the iron and corresponds to the first part of the magnetization curve of Fig. 5.1 where the slope is small.

As the field is increased, the stage corresponding to the steep part of the magnetization curve is reached. Here the domains change suddenly

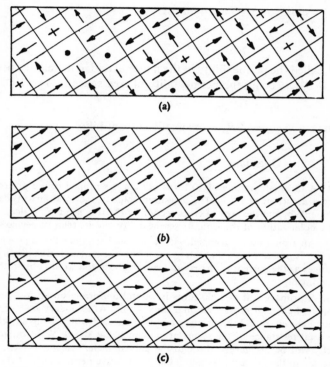

Fig. 5.16. (a) Unmagnetized. (b) Partially magnetized. (c) Magnetically saturated.

the direction of their magnetic axes to that of one of the six natural, easy directions most nearly in the direction of the field (Fig. 5.16 (b)).

Finally, at the top of the magnetization curve, where the slope is again small, the axes of the domains are gradually rotated into the direction of the field (Fig. 5.16 (c)).

Hysteresis loss is caused by small eddy currents which are induced by the sudden orientations of the domains.

Summary

$$B = B_0 + J.$$

J is the magnetic polarization of a material and it is equal to the magnetic flux density provided by the material; it is also equal to the electromagnetic moment per unit volume. Its units are webers per square metre and it is a vector quantity.

$$\mu_r = \frac{\mu}{\mu_0}.$$

μ is called the permeability of the material and μ_r its relative permeability $= B/B_0$, μ_0 being the permeability of free space $= 4\pi \times 10^{-7}$ H m^{-1}.

$$\mu_r = 1 + \chi_m$$

where $\chi_m = J/B_0$ = magnetic susceptibility.

Questions

1. A toroidal winding of 10 turns per cm is wrapped round an iron ring and carries a current of 1·5 A. The measured value of the total flux inside the ring is 0·060 mWb, and the area of cross-section of the ring is 3 cm². Calculate the magnetic polarization of the iron, its permeability and its relative permeability.

2. An iron ring is wrapped with a toroidal winding through which a small current is passed. Draw a graph to show how the magnetic flux density in the ring changes as the current is increased. Describe how you would obtain this graph by experiment.

3. A primary coil of 1200 turns is wound uniformly on an iron ring of mean radius 14 cm and cross-secitional area 4 cm². A secondary coil of 50 turns is wound closely round the primary and is connected with a ballistic galvanometer, the total resistance of the secondary circuit being 300 Ω. What quantity of electricity will be discharged through the galvanometer when a current of 3·0 A is reversed in the primary, assuming that the relative permeability of the iron under these conditions is 500?

4. A long solenoid, having 10 turns per cm length, has an iron core of area 4·00 cm² and relative permeability 100, and carries a current of 2·00 A. Calculate the quantity of electricity induced in a secondary, of 100 turns and total resistance of 2·00 Ω, wound round the outside of the solenoid, when the iron core is withdrawn.

5. Explain the meanings of *hysteresis* and *remanence*. How do the magnitudes of these quantities for a sample of iron determine the magnetic purposes for which it is suitable? *(Contd. over.)*

Describe how you would obtain by experiment the hysteresis curve for iron in the form of a ring.

6. When a bundle of iron wires is placed in a coil of wire carrying a.c. a rise in temperature of the iron is observed. Explain the causes of this heating and show how the magnitude of the effect depends on (*a*) the magnetic properties of the iron, (*b*) the electrical conductivity of the iron, (*c*) the frequency of the a.c., (*d*) the diameter of the iron wires forming the bundle.   (CS)

7. When the iron yoke in Fig. 5.17 is lowered to close the iron circuit, an induced current flows in the galvanometer. Explain why this occurs and what is the source of the energy.

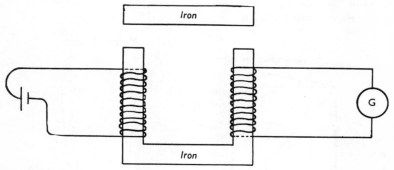

Fig. 5.17

8. What is meant by a *ferromagnetic material*? Describe the domain theory of ferromagnetism and explain how it accounts for the shape of the magnetization curve.

9. Iron, cobalt and nickel are said to be *ferromagnetic*, oxygen and green vitriol ($FeSO_4 . 7H_2O$) *paramagnetic*, bismuth *diamagnetic*. Describe what is meant by this statement and explain as far as you can one of these phenomena.   (OS)

10. Give a general account of the magnetization of iron. Show how the processes of magnetization and demagnetization, the phenomenon of hysteresis, and the existence of a Curie temperature can be explained in terms of elementary magnets and a domain structure.

What magnetic properties are desirable for the material of

    (*a*) the core of a transformer,

    (*b*) the core of a relay electromagnet,

    (*c*) the tape of a magnetic tape recorder?   (O 1970)

6   Terrestrial magnetism

A magnetic needle, when suspended freely so that it can turn in both a horizontal and a vertical plane, sets itself, in Britain at the present time, in a vertical plane about 9° W of N with its N pole dipping about 67° below the horizontal; it follows that the magnetic field due to the earth must lie in this direction.

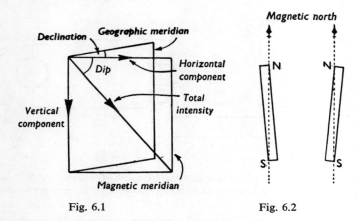

Fig. 6.1   Fig. 6.2

The following terms are in common use (see Fig. 6.1):

The geographic meridian is the vertical plane in a direction geographic N and S, i.e. which passes through the earth's geographic poles.

The magnetic meridian is the vertical plane in which a magnet sets itself at a particular place.

The angle of declination (or variation of the compass) is the angle between the magnetic and geographic meridians.

The angle of dip (or inclination) is the angle between the horizontal and the magnetic axis of a magnet free to swing in the magnetic meridian about a horizontal axis. It is the angle between the directions of the earth's magnetic field and the horizontal.

The magnetic field of the earth, often called the *total intensity*, is resolved for convenience into a *horizontal component* and a *vertical component*.

The horizontal component of the magnetic flux density was about 1.9×10^{-5} Wb m^{-2} in Britain in 1966.

The quantities, declination, dip, total intensity, horizontal component and vertical component, are known as the *magnetic elements*. The earth's magnetic field at a particular place may be specified by the declination and any two of the other magnetic elements.

Determination of declination

The determination of the angle of declination at a place involves finding two directions, geographic N and magnetic N.

The former can be found accurately only by an astronomical method – observation of the sun or stars. It can be found with fair accuracy from the fact that the shadow of a vertical stick cast by the sun at mid-day is due N.

Magnetic N is found by suspending a bar magnet freely on a vertical axis. Since the magnetic axis of the magnet may not coincide with its geometric axis, the magnet must be turned over and the mean of its two directions found (Fig. 6.2).

The earth inductor

Fig. 6.3

The horizontal and vertical components of the earth's magnetic flux density can be measured by the earth inductor (Fig. 6.3). The instrument consists of a coil of wire which can be rotated about an axis capable of being set in any direction by turning a movable frame. The principle of the instrument is that the coil is turned through a right angle between positions when it is threaded by maximum magnetic flux and zero magnetic flux, and the quantity of electricity induced is measured by a ballistic galvanometer connected in series with the coil.

The quantity of electricity induced is given by

$$Q = -\frac{N\Phi}{R} \quad \text{(p. 83)},$$

where N is the number of turns in the coil, Φ is the change in the magnetic flux threading it and R is the total resistance of the coil and ballistic galvanometer circuit.

Suppose that the coil is perpendicular to a field of magnetic flux density B and is then turned through a right angle so that no magnetic flux threads it.

$$\Phi = BA,$$

where A is the area of the coil.

$$\therefore \ B = -\frac{QR}{NA}.$$

Thus, measuring Q, R and A, and knowing N, B can be calculated.

To determine the horizontal component of the earth's magnetic flux density, the frame is made vertical and the whole instrument set magnetic E and W. In this position the coil is perpendicular to the earth's magnetic field and the maximum horizontal magnetic flux threads it. The coil is turned through 90° (about its vertical axis) so that its plane then lies in the magnetic meridian and none of the earth's magnetic flux threads it.

To determine the vertical component of the earth's magnetic field the frame is turned so that the coil can move about a horizontal axis in the magnetic meridian. When the coil is rotated through 90°, say from a vertical to a horizontal position, the quantity of electricity induced is proportional to the vertical component.

Determination of dip

The angle of dip, D, can be calculated from the values of the horizontal and vertical components:

$$\tan D = \frac{\text{vertical component}}{\text{horizontal component}} = \frac{\theta_v}{\theta_h},$$

where θ_v and θ_h are the throws of the ballistic galvanometer in the earth inductor experiment.

Alternatively, the direction of the axis of the coil of the earth inductor can be adjusted until no current flows when the coil is rotated. The direction of the axis is then the direction of the resultant earth's magnetic field.

The deflection magnetometer

Two magnetic fields can be compared by means of a deflection magnetometer, which consists of a small magnet, pivoted on a vertical axis and carrying a light pointer which can move over a circular scale.

Normally one of the fields is the earth's horizontal component and the other field is arranged to be at right angles to this. The pivoted magnet sets itself along the resultant of the two fields, at an angle θ to its direction when it is in the earth's field alone. If B_e is the magnetic flux density of the earth's horizontal component and B is the magnetic flux density of the other field,

$$B = B_e \tan \theta \quad \text{(Fig. 6.4).}$$

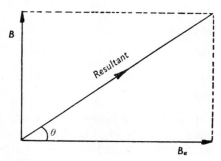

Fig. 6.4

Measurement of the horizontal component of the earth's magnetic flux density using a deflection magnetometer

The magnetic flux density at the centre of a circular coil of known radius and known number of turns, when a measured current is passing through the coil, can be calculated. This can be compared, by means of a deflection magnetometer, with the horizontal component of the earth's magnetic flux density, enabling the latter to be determined.

A convenient instrument for the purpose is a tangent galvanometer, which consists of a circular coil, at the centre of which there is a

Fig. 6.5

deflection magnetometer (Fig. 6.5). If N is the number of turns in the coil, a its radius, I the current, and B the magnetic flux density at the centre of the coil,

$$B = \frac{\mu_0 NI}{2a}$$

Using the same nomenclature as above,

$$B = B_e \tan \theta$$

$$\therefore \ B_e = \frac{\mu_0 NI}{2a \tan \theta}$$

The tangent galvanometer is connected through a reversing switch to a battery, rheostat and accurate ammeter (Fig. 6.6). Corresponding values of I, as measured by the ammeter, and θ are taken. A graph of I and $\tan \theta$ can be plotted, which will be a straight line of slope $I/\tan \theta$. Knowing also N and a, B can be calculated.

The reversing switch enables the current to be reversed in the galvanometer coil without changing the direction of the current in the rest of the circuit. By taking readings of the deflection when the current is flowing in opposite directions through the coil, various errors, for example, that due to imperfect setting of the coil in the magnetic meridian, may be eliminated.

EXAMPLE

A tangent galvanometer has a coil of 2 turns of mean radius 7·5 cm, which is set with its plane in the magnetic meridian. A current is passed through the coil and

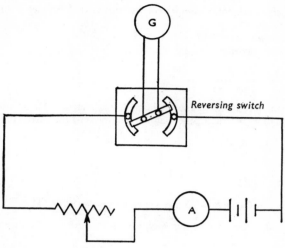

Fig. 6.6

produces a deflection of a magnet, pivoted at the centre of the coil, of 45°. Calculate the current if the horizontal component of the earth's flux density is 1.8×10^{-5} Wb m^{-2}.

Let I = current in the coil.
Magnetic flux density at centre of coil,

$$B = \frac{\mu_0 NI}{2a}$$

$$= \frac{2I\mu_0}{2 \times 7.5 \times 10^{-2}}.$$

Magnetic flux density due to the earth = 1.8×10^{-5}.

$$B = B_e \tan \theta$$

$$\therefore \frac{2I\mu_0}{2 \times 7.5 \times 10^{-2}} = 1.8 \times 10^{-5} \tan 45°,$$

$$I = 1.07 \text{ A}.$$

EXAMPLE

A circular coil having 50 turns of mean radius 8.0 cm is set with its coil in the magnetic meridian and a small magnet is pivoted at its centre. When a current is passed through the coil it is found that the coil must be rotated through 40° before the magnet is once again in the plane of the coil. Calculate the current in the coil if the horizontal component of the earth's magnetic flux density is $1 \cdot 8 \times 10^{-5}$ Wb m^{-2}.

Let I = current in the coil.

Magnetic flux density at centre of coil, $B = \dfrac{NI\mu_0}{2a}$

$$= \frac{50I\mu_0}{2 \times 8 \cdot 0 \times 10^{-2}}.$$

B is at right angles to the plane of the coil. The magnet will set itself along the resultant, R, of B and of B_e, the magnetizing force of the earth (Fig. 6.7).

$$B = B_e \sin 40°$$

$$\therefore \frac{50I\mu_0}{2 \times 8 \cdot 0 \times 10^{-2}} = 1 \cdot 8 \times 10^{-5} \sin 40°,$$

$$I = 0 \cdot 029 \text{ A} = 29 \text{ mA} \quad \text{(using } \mu_0 = 4\pi \times 10^{-7} \text{ H m}^{-1}\text{).}$$

The instrument, when used in this way, is called a sine galvanometer.

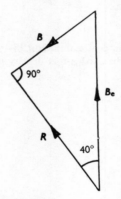

Fig. 6.7

Variation of dip over the earth's surface

The angle of dip is 0° approximately at the equator. It increases steadily northward or southward, until it becomes 90° at the magnetic poles. In the northern hemisphere the N pole dips, and in the southern hemisphere the S pole dips. Fig. 6.8 is a map showing isoclinic lines, which are lines joining places at which the angle of dip is the same.

Gilbert constructed a model spherical magnet of lodestone, which he called a terrella or 'little earth', with poles as shown in Fig. 6.9. By means of this model he was able to show that the angle of dip should vary over the surface of the earth as described. The dotted lines in the

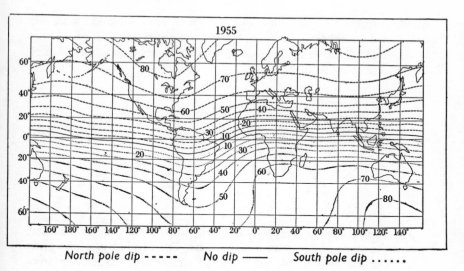

North pole dip - - - - -   No dip ——   South pole dip ......

Fig. 6.8

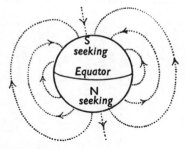

Fig. 6.9

figure represent the lines of magnetic flux, and the arrows represent compass needles, the arrowheads being the N poles.

The angle of dip at the magnetic poles is 90°; it is in this way that they are located. The N magnetic pole was discovered in the far north of Canada in 1831 by Sir James Ross; aircraft flights over the pole have shown that its position has shifted by some hundreds of miles. The S magnetic pole was discovered by Sir Ernest Shackleton in 1909.

Variation of declination over the earth's surface

Since the geographic and magnetic poles do not coincide, the declination varies over the earth's surface. Fig. 6.10 shows the great circle drawn through all four poles; this is only an approximation since the four poles do not lie exactly on a great circle. The declination on this circle should be zero; on one side of it the declination should be E of N and on the other W of N.

Fig. 6.10

Figure 6.11 shows isogonic lines, which are lines joining places at which the angle of declination is the same. The two halves of the great circle can be recognized, but one-half describes a large loop over Siberia, due to the irregular magnetization of the earth's crust.

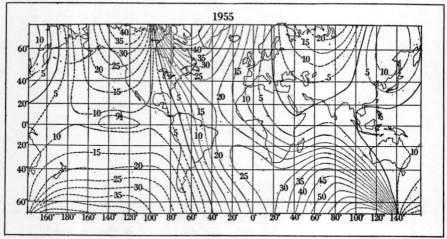

Easterly variation - - - - -   True north ——   Westerly variation . . . . . .

Fig. 6.11

For many years in the seventeenth century it was hoped that changes in magnetic declination might be the clue to an accurate method of finding longitude, an urgent navigational problem for the solution of which the British Government offered a prize of £20 000. In 1617 Henry

Bond expounded a method based on magnetic declination in a book, *Longitude Found*, answered two years later by Peter Blackbarrow in *The Longitude Not Found*. The problem was solved not by a magnetic method but by the development of a suitable chronometer by Harrison.

Christopher Columbus is sometimes credited with the discovery of the change in declination when he sailed across the Atlantic to the Bahamas in 1492. It was probably known, however, to earlier navigators. Some of the portable sundials carried by travellers in the fifteenth century as time-pieces, had a compass in the base for setting, and a mark indicating the amount of declination. Edmund Halley, between October 1698 and September 1700, made two voyages in the North and South Atlantic, the first sea voyages made for purely scientific purposes, which resulted in the first declination chart of the earth. There are now about seventy-five magnetic observatories distributed over the surface of the earth. In 1882–3 and 1932–3 many nations combined to set up magnetic observatories in the Arctic regions and results of great value were obtained.

Secular variations

Besides varying in space over the surface of the earth, the magnetic elements are slowly varying in time. Measurements of the angle of declination in Greenwich have been made since the year 1580. The following table shows the variation from that date until 1966:

1580	11° E	1900	17° W
1622	6° E	1946	9° 51′ W*
1658	0°	1955	8° 44′ W*
1816	24° W	1966	9·417° W

The angle of dip is also varying, but to a smaller extent than the declination. In 1580 its value in London was 72°, in 1658 74°, in 1816 71°, in 1946 66° 44′,* and in 1955 66° 39′.*

Maps to illustrate the secular variations, for example lines joining places at which the horizontal or vertical component of the earth's magnetic field is changing at the same rate, bear a remarkable resemblance to weather charts, with a number of centres of disturbance.

Besides these slow secular changes, as they are called, there occur at intervals magnetic storms, when the needles of all magnetic recording instruments are affected in a comparatively violent manner. Magnetic

* The values for 1946 and 1955 are those at Abinger, near Dorking, where the magnetic elements were then recorded. Because of disturbances due to electric traction, the Observatory was moved in 1956 from Abinger to Hartland, north Devon.

storms have been traced to electrons in the upper atmosphere, originating from sunspots. These are also responsible for the Aurora Borealis or Northern Lights, a beautiful and spectacular glow to be seen occasionally in the sky in northern latitudes. Hiorter, an observer in Upsala, who was one of the pioneers in this sphere, wrote in 1747: 'Who could have thought that the northern lights would have a connexion and sympathy with the magnet, and that these northern lights, when they draw southwards across our zenith or descend unequally towards the eastern and western horizons, could within a few minutes cause considerable oscillations of the magnetic needle through whole degrees?'

Magnetic storms show an 11-year cycle of activity which corresponds with the 11-year sunspot cycle; they also tend to recur every 27·3 days, the period of rotation of the sun.

Magnetic rocks

Records of the earth's magnetic field have been kept for only about 400 years, but new evidence has recently been found which may provide knowledge of the secular variations during the past 500 million years. Certain rocks, which contain traces of iron oxide minerals such as haematite (Fe_2O_3) and magnetite (Fe_3O_4), are weakly magnetized. These rocks are of two types, igneous and sedimentary. The former include lavas, which poured from volcanoes millions of years ago, and became magnetized as they cooled, in the direction of the earth's magnetic field of that time; the latter include sandstones and shales, formed of compressed and hardened deposits of sand and mud, the direction of whose magnetization was determined also when the rocks were formed.

Lavas in several parts of the world, such as those of Mount Etna in Sicily, of the Auvergne in France, of south-west Iceland, and of the Columbia River, USA, have been investigated. Each of these consists of a series of flows which occurred over a vast period of time, and it has been found that about half the flows are magnetized in roughly the present direction of the earth's magnetic field, while the other half are magnetized in the opposite direction. Hospers has deduced that the direction of the earth's magnetism has been reversed about every million years. Similar evidence has been obtained from the new red sandstones of England.

There is, however, a conflict of opinion about the reversal of the earth's magnetic field. Nagata in Japan has shown that certain specimens of pumice from Mount Haruna can acquire a reverse magnetization on cooling in a magnetic field. It is possible, therefore, that the reversed magnetism of rocks may be due to self-reversing minerals rather than to a reversed field of the earth.

The magnetism of rocks may also provide decisive evidence on the theory of continental drift. The west coast of Africa and the east coast of South America have similar shapes, and it has been suggested that the two continents were once joined and have drifted apart. According to a well-known theory of Wegener, Australia, India, South Africa and the southern part of South America were once joined to Antarctica and have drifted northwards. Since the angle of dip corresponds very closely to latitude, it is possible to determine, from the inclination of the magnetic axes of the rocks, in what latitude they became magnetized.

Much work is now being done in various parts of the world on the problems associated with the magnetism of rocks.

The origin of the earth's magnetism

The main features of the earth's magnetic field can be accounted for by assuming that the earth's crust is a comparatively uniformly magnetized shell with some large irregularities. The shell cannot be thicker than 20 kilometres, since at depths greater than this the earth is too hot to be magnetic. Measurements indicate, however, that the intensity of magnetization of the crust is insufficiently high to account for the field.

The sun has a magnetic field similar to that of the earth, and its magnetic moment bears the same ratio to its angular momentum as in the case of the earth. Professor Blackett has suggested that every rotating body creates a magnetic field by virtue of its rotation. In 1947 the magnetic field of a star, 78 Virginis, was measured by the American astronomer, Babcock, by means of its effect on the spectral lines, i.e. the Zeeman effect, and the magnetic moment of the star bore the predicted relation to the angular momentum. But Babcock has since found several stars with magnetic fields which are varying and even changing direction; it seems unlikely that the rotation of a star accelerates and reverses.

Blackett's theory has not been put to a direct experimental test because a large mass, whirled at a very high speed, would be needed to produce even a small magnetic field. The theory requires that the horizontal component, but not the vertical component, of the earth's field should decrease with depth, and measurements have been made at the bottom of mines. These have been inconclusive because no mine is deep enough to produce an effect which is not masked by disturbances from small amounts of ferromagnetic materials in the local rocks. A measurement of the magnetic field of the moon would be an admirable test of the theory; astronauts *landing* on the moon have found a very small field.

Another theory, developed by the American, W. M. Elsasser, and by Sir Edward Bullard, is that the earth behaves like a self-exciting dynamo, creating its magnetic field by means of electric currents induced by that field in the core, and deriving its energy from the heat developed by radioactive elements in the core. The core has a radius about half that of the earth and is believed to consist mainly of molten iron, which is a much better electrical conductor than the siliceous material of the surrounding shell or mantle.

A simple, self-exciting dynamo, which serves to illustrate the underlying idea of the Elsasser–Bullard theory, is shown in Fig. 6.12. A metal disk is rotated and its rim and axle are connected by rubbing contacts

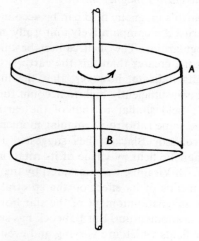

Fig. 6.12

at *A* and *B* to a stationary, coaxial coil. If there is a small initial vertical magnetic field, an e.m.f. will be induced between the rim and the axle of the disk, and this will send a current through the coil. The current in the coil will produce a magnetic field which reinforces the initial magnetic field (so long as the rotation of the disk is in the correct direction), and the field will grow until equilibrium is attained when the energy dissipated in heat equals that supplied to the rotating shaft.

The heat generated by radioactive elements in the earth's core, it is suggested, gives rise to convection currents, the material of the core rising in some places and sinking in others. The sinking material carries angular momentum to the centre of the core and accelerates the rotation there, while the rising material causes a corresponding deceleration of the rim. The centre of the core, therefore, rotates more quickly than the

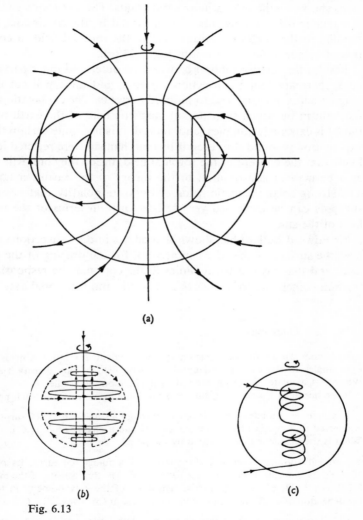

(a)

(b)

(c)

Fig. 6.13

rim, just as water in a wash basin, running into a central drain, rotates more quickly near the drain.

Let Fig. 6.13 (a) represent the earth's magnetic field, and imagine these lines of magnetic flux to be running through the core in Fig. 6.13 (*b*), parallel to the axis of rotation. The core will cut these lines of magnetic flux as it rotates, the outer parts of the core cutting them more slowly than the inner parts, and hence induced currents will flow in the directions shown by the discontinuous quadrants. These currents give rise to a magnetic field (among others) represented by the continuous circles, parallel to the circles of latitude, called the toroidal field; a complete line of magnetic flux is shown in Fig. 6.13 (*c*).

Besides the relative motion between the outer and inner parts of the core, there are radial convection currents, and these will cut lines of magnetic flux and give rise to induced currents. The explanation of the mechanism by which the external magnetic field of the earth is maintained is extremely complex, and, indeed, it is not quite certain that the explanation is sound. But even if this mechanism were rejected it seems likely that the concept of the toroidal field, inside the core, will remain as a permanent contribution to the theory of the earth's magnetism. The strong solar magnetic fields of opposite polarity that occur with sunspots can be explained very satisfactorily in terms of the toroidal field of the sun.

Sir Edward Bullard has shown that eddies in the convection currents near the surface of the core can account for the drifting of the earth's field, and that asymmetrical eddies in the core may be responsible for the non-coincidence of the earth's magnetic and rotational axes.

Questions

1. A compass needle of electro-magnetic moment $2 \cdot 0 \times 10^{-6}$ A m^2 is held at right angles to the earth's horizontal magnetic field of flux density $1 \cdot 8 \times 10^{-5}$ Wb m^{-2}. Calculate the torque exerted by the field on the needle.
If the needle is $1 \cdot 5$ cm long, calculate the force in Newtons on each pole.

2. One pole of a bar magnet is passed through a small coil of 100 turns of wire connected in a circuit of total resistance 5 Ω and a charge of 200 μC is induced. What is the magnetic flux provided by the pole?

3. A galvanometer coil is 2 cm square and contains 100 turns. Its period is 20 s and its moment of inertia is 2×10^{-7} kg m^2. The flux density of the magnet is $0 \cdot 2$ Wb m^{-2}. Determine the current sensitivity of the galvanometer in terms of the deflection of a reflected beam of light on a screen 1 m away.   (CS adapted.)

4. A circular coil of 50 turns of mean radius 10 cm stands in the magnetic meridian, and a small magnetic needle at its centre is deflected through 60°. Assuming the horizontal component of the earth's flux density to be $2 \cdot 0 \times 10^{-5}$ Wb m^{-2}, calculate the current in the coil.

5. In reading the deflection of a tangent galvanometer, it is usual to read both ends of the pointer for both directions of the current. Explain which of the following errors are thereby eliminated: (i) inaccurate setting of the coil in the magnetic meridian, (ii) bent pointer, (iii) pointer not in the centre of the circular scale, (iv) centre of the circular scale not in the centre of the coil, (v) line joining the zeros of the circular scale not at right angles to the plane of the coil.

6. A circular coil of 50 turns and mean radius 7·0 cm is set with its plane in the magnetic meridian. There is a small pivoted magnetic needle at its centre. When a current of 0·021 A is passed through the coil it is found that the coil must be rotated through 30° before the magnet is once again in the plane of the coil. Find the horizontal component of the earth's magnetic flux density.

7. Describe in detail how the earth inductor may be used to determine the angle of dip.

The coil of an earth inductor, connected to a ballistic galvanometer, is rotated through 90° from a position in which it is threaded by the horizontal component of the earth's field to a position in which it is not threaded, and the throw of the galvanometer is 25·0 divisions. When it is rotated through 90° between positions in which it is threaded, and not threaded, by the vertical component of the earth's field the throw is 57·5 divisions. What is the angle of dip?

8. A circular coil of 500 turns and radius 10·0 cm stands with its plane vertical and facing magnetic north and south. Its resistance is 50 Ω and it is connected to a ballistic galvanometer of resistance 150 Ω. When it is suddenly turned through 180° about a vertical axis, the throw of the galvonometer is 3·00 μC. What is the horizontal component of the earth's magnetic flux density?

9. A ballistic galvanometer is connected in series with an earth inductor and with a coil of 50 turns, wound over a long solenoid having 10 turns per cm and of radius 5 cm. The earth inductor has 200 turns, of mean area 1000 cm². It is found that the maximum deflection obtainable by a 180° rotation of the earth inductor is the same as the deflection obtained by reversing a current of 2·00 A in the solenoid. Find the resultant flux density of the earth's magnetic field.

10. Give an account of the way in which the earth's magnetism varies (a) in place and (b) in time.   (O & C)

7 Electrolysis and cells

Most liquids which conduct electricity are split up chemically by the current. This process is called electrolysis and the liquids are called electrolytes. Most solutions of acids, salts and bases are electrolytes; the commonest liquid conductor which is not an electrolyte is mercury.

The electrode by which the flow of electrons enters the electrolyte is called the cathode, and that by which the flow of electrons leaves, the anode. The same names are applied to the thermionic diode (see Chapter 10), but here the similarity stops, because in a thermionic diode it is electrons which constitute the charge flow from cathode to anode, whereas in electrolysis the charge carriers are ions. The process of copper plating will be described by way of example.

When a current is passed between copper electrodes through an aqueous solution of copper sulphate, it is found that copper is removed from the anode and deposited on the cathode. It is assumed that the copper sulphate molecules in water become copper ions with a ' + ' charge (shortage of electrons), and sulphate ions with a ' − ' charge:

$$CuSO_4 \rightarrow Cu^{2+} + SO_4^{2-}.$$

The external electricity supply is taking electrons from the anode, so the SO_4^{2-} ion is attracted towards it. At the anode it gives up its two electrons, at the same time combining with part of the anode copper to produce more copper sulphate:

$$Cu + SO_4 \rightarrow CuSO_4.$$

This is called a 'secondary reaction'. Meanwhile the Cu^{2+} ions are attracted to the cathode, which is being supplied with electrons by the external electricity supply. Two such electrons will neutralize a Cu^{2+} ion, leaving metallic copper which adheres to the cathode.

If platinum electrodes are used, it is found that the SO_4^{2-} ions do not combine with the anode; instead the SO_4 radical combines with some of the water:

$$2SO_4 + 2H_2O \rightarrow 2H_2SO_4 + O_2.$$

Oxygen is seen bubbling from the anode, and an indicator will confirm the generation of an acid in the vicinity of the anode.

The idea of secondary reactions has now been replaced by the theory of discharge potentials. But, for simple electrolytic phenomena, secondary reactions provide a satisfactory working hypothesis and do not put such a tax on the memory as discharge potentials.

Faraday's laws of electrolysis

The quantitative investigation of electrolysis was performed by Faraday who summarized his results in two laws:

(1) *The mass of a substance liberated in electrolysis is proportional to the quantity of electricity passed, i.e. to the product of current and time.*

(2) *When the same quantity of electricity is passed through different electrolytes the masses of substances liberated are in the ratio of their equivalent weights.**

The quantity of electricity which liberates 1 mole of all singly-charged ions, $\frac{1}{2}$ mole of doubly-charged ions, and so on, is approximately 96 500 coulombs and is called the *faraday*. The faraday is not an SI unit.

The electrochemical equivalent

The electrochemical equivalent of a substance is the mass liberated in electrolysis by 1 coulomb.

This is a quantity which can be determined accurately by experiment and from which the faraday may be calculated:

$$\text{faraday} = \frac{\text{equivalent weight of substance}}{\text{electrochemical equivalent of substance}}.$$

EXAMPLE

A current of 0·800 A is passed for 30 minutes through a water voltameter, consisting of platinum electrodes dipping into dilute sulphuric acid, and it is found that 178·0 cm³ of hydrogen at a pressure of 75·0 cm of mercury at 288 K are liberated. Calculate the electrochemical equivalent of hydrogen, given that its density at s.t.p. is $8 \cdot 99 \times 10^{-2}$ kg m^{-3}.

If a copper voltameter, consisting of copper plates dipping into a solution of copper sulphate, had been connected in series with the water voltameter, calculate the mass of copper deposited on the cathode, given that the equivalent weights of hydrogen and copper are 1 and 31·8 respectively.

$$\text{Volume of hydrogen at s.t.p.} = 178 \times \frac{75}{76} \times \frac{273}{273 + 15}$$
$$= 166 \cdot 5 \text{ cm}^3.$$
$$\text{Mass of hydrogen} = 166 \cdot 5 \times 8 \cdot 99 \times 10^{-2} \times 10^{-6}$$
$$= 1 \cdot 50 \times 10^{-5} \text{ kg}.$$

* Relative atomic mass ÷ charge on each iron.

$$\text{Charge of electricity passed} = 0.800 \times 30 \times 60$$
$$= 1440 \text{ C}.$$

$$\therefore \text{ Electrochemical equivalent of hydrogen} = \frac{1 \cdot 50 \times 10^{-5}}{1440}$$
$$= 1 \cdot 04 \times 10^{-8} \text{ kg C}^{-1}.$$

$$\frac{\text{Mass of copper deposited}}{\text{Mass of hydrogen liberated}} = \frac{31 \cdot 8}{1}$$

$$\therefore \text{ Mass of copper deposited} = 31 \cdot 8 \times 1 \cdot 50 \times 10^{-5}$$
$$= 4 \cdot 77 \times 10^{-4} \text{ kg}.$$

Theories of electrolysis

A theory of electrolysis must account for the way in which an electrolyte carries a current and how the electrolyte is decomposed.

It was very early suggested that the molecule of an electrolyte must consist of positively and negatively charged parts and that the positive part is attached to the cathode and the negative part to the anode.

When it was discovered that Ohm's law holds for electrolytes and that, if polarization is avoided, the most minute e.m.f. will cause some current to flow, it was realized that part of the e.m.f. could not be employed in decomposing the electrolyte. Clausius, in 1857, suggested that some of the electrolyte might be split up in solution into positively and negatively charged particles before an e.m.f. was applied. Faraday gave the name *ions* to these charged particles.

The theory of Arrhenius or ionic theory

In 1887 Arrhenius put forward his theory of electrolytic dissociation. He assumed that when an electrolyte enters solution part of it is dissociated into ions. Thus when sodium chloride, NaCl, is dissolved in water some of the molecules dissociate into positively charged sodium ions, Na^+, and negatively charged chlorine ions, Cl^-. Ions are continually recombining and molecules dissociating, there being a dynamic equilibrium represented by the equation

$$NaCl \rightleftharpoons Na^+ + Cl^-.$$

The theory met with violent opposition at first. Sodium and chlorine are extremely active chemically, and their separate existence in a solution was incredible to many chemists; it was not realized that ions have very different chemical properties from the uncharged atoms. We can now understand clearly how the sodium and chlorine ions gain their charges and why their chemical properties differ from those of the corresponding uncharged atoms. A sodium atom has a nucleus, with a positive charge of 11, and 11 electrons round the nucleus. It will readily part with one electron, leaving it with a resultant positive charge and with the very stable arrangement of 10 electrons possessed by its predecessor in the periodic table, the inert gas neon. Similarly, a chlorine atom readily gains an electron, obtaining a negative charge and increasing its 17 electrons to the stable configuration of 18 electrons, possessed by the inert gas argon.

When an e.m.f. is applied between electrodes dipping into an electrolyte the positive ions, or cations, are attracted to the cathode while the negative ions, or anions, are attracted to the anode. The two streams of oppositely charged ions, travelling in opposite directions, carry the current through the electrolyte. Anions give up their surplus electrons to the anode, and cations receive electrons from the cathode, thus maintaining the flow of electrons in the external circuit. Having given up their charges the ions are liberated as uncharged atoms and molecules.

The charge on the ion

If e is the charge on the hydrogen ion and N_A is the number of ions in 1 g of hydrogen,

$$N_A e = 96\,500$$

since 1 g of hydrogen is liberated by 96 500 coulombs. N_A is known as the Avogadro Constant and may be determined from the kinetic theory of gases; its value is $6 \cdot 06 \times 10^{23}$.

$$\therefore \ e = \frac{96\,500}{6 \cdot 06 \times 10^{23}}$$

$$= 1 \cdot 59 \times 10^{-19} \text{ C.}$$

This agrees with the value for the charge of the electron determined by other methods described in Chapter 13.

In the case of a dipositive substance such as copper, a mole* contains N_A atoms, but, since 96 500 coulombs discharges only $\frac{1}{2} N_A$ ions of a 2-charge substance, it must be assumed that each ion carries a charge $2e$. Thus

$$CuSO_4 \rightarrow Cu^{2+} + SO_4^{2-}.$$

* A mole is the amount of substance which contains as many elementary units as there are atoms in 12 g of carbon-12.

The ions of any substance have a surplus or deficiency of electrons equal to a whole number of electrons.

In this way Faraday's second law can be accounted for on the ionic theory.

Evidence for dissociation

There is strong evidence of the dissociation of electrolytes in solution quite apart from electrolysis. The particles of a solute may be considered as distributed in a solution rather like the molecules of a gas. They set up a pressure, called the osmotic pressure, which can be measured in some cases with the help of a membrane, which is permeable only to the solvent molecules. For very dilute solutions the osmotic pressure is proportional to the concentration and to the absolute temperature. Van't Hoff showed theoretically that the gas laws can be applied to the solute in very dilute solutions. He found that the gas equation $pv = RT$ holds for non-electrolytes such as cane sugar. But the osmotic pressure of electrolytes is much greater than would be expected from a consideration of the number of molecules; in the case of sodium chloride it is nearly twice as great. This can be explained by assuming that each molecule of sodium chloride has dissociated into two ions, and hence there are twice as many particles contributing to the osmotic pressure. Similarly, the depression of the freezing-point and elevation of the boiling-point of solutions of electrolytes is abnormally large owing to dissociation.

Again, solutions of electrolytes exhibit the characteristics of their individual ions. All acids have a sour taste which is ascribed to the hydrogen ion, all cupric salts are blue due to the cupric ion, all dichromates are yellow due to the dichromate ion, and so on.

Degree of dissociation

The electrical conductivity of a solution containing a given mass of an electrolyte increases with increasing dilution and ventually approaches a maximum value. Arrhenius explained this by saying that on dilution the electrolyte becomes more dissociated; at infinite dilution it is completely dissociated.

The theory of Debye and Hückel

It is now known, as a result of the investigation of crystal structure with X-rays, that the crystals of many compounds are completely ionized in the solid state. Fig. 13.16 on page 303 represents a crystal of rock salt, NaCl; an electron has been transferred from each of the sodium atoms to a chlorine atom, and the cubic lattice is composed solely of sodium and chlorine ions, held in position by their electrostatic attraction.

Debye and Hückel put forward in 1923 the theory that substances such as rock-salt, being completely ionized in the solid state, are also completely ionized in solution at all concentrations. How, then, can the marked change in conductivity with dilution be explained?

Arrhenius had assumed that an increase in conductivity is caused solely by an increase in the number of ions present in the solution; he believed that the mobilities of the ions do not vary as a solution is diluted. Debye and Hückel, on the other hand, started by assuming that the number of ions of strong electrolytes is constant at all concentrations and accounted for the change in conductivity on dilution by a change in the mobility of the ions.

Cells

Cells in use nowadays are divided into two groups: unrechargeable (primary) and rechargeable (secondary).

Primary cells

The commonest 'dry' cell is the dry Leclanché (Fig. 7.1 (a)). It consists of a carbon rod (positive or electron-receiving terminal), surrounded by a paste of magnesium dioxide. Outside this is a concentric paste of ammonium chloride (the electrolyte), contained in a zinc case which is also the negative terminal. The chemical reactions which take place can be represented by the equation

$$Zn + 2NH_4Cl + 2MnO_2 \rightarrow Zn(NH_3)_2Cl_2 + H_2O + Mn_2O_3.$$

The manganese dioxide is referred to as the 'depolarizer', because without it bubbles of hydrogen would form on the carbon rod, forming an insulating layer around it. With the MnO_2 paste, this hydrogen is at once converted into water, which does not interfere with the cell's action.

Hydrogen ions give up their + charge (i.e. acquire a neutralizing electron) at the carbon rod. When the ammonium chloride reacts with the zinc case, the latter is left with an extra electron. Thus the electron flow in the external circuit is from zinc (− terminal) to carbon (+ terminal). The layer form is shown in Fig. 7.1 (b).

The e.m.f. of the cell is about 1·5 volts; in use, its internal resistance rises until it is of no further use. However, although the cell cannot be 'recharged' in the sense of completely reversing the chemical reactions,

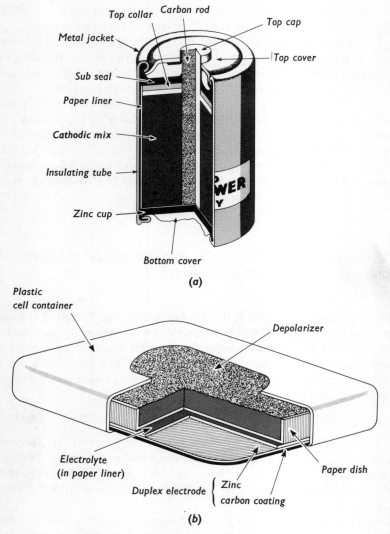

Fig. 7.1. Cross-sections of (*a*) Round cell, and (*b*) Layer cell.

it can be 'rejuvenated' – a process which goes part of the way. Partly-rectified a.c. is used, so that in a cycle it is slightly charged, and then slightly *less* discharged. By this process its life can be extended to about five times its first life, in some ten rejuvenating processes.

One of the main disadvantages of the dry Leclanché cell lies in the fact that the zinc case takes part in the reaction, and is therefore slowly eaten away. Eventually it will puncture, and the corrosive electrolyte will spill out, endangering the equipment it is supposed to be energizing. 'Leakproof' cells have an extra outer covering, but are only relatively safe in this respect.

The mercury cell

As the mercury cell has been tailored to so many purposes and sizes, it is most easily explained by reference to its original form (Fig. 7.2). In this cell the chemical reaction is

$$Zn + HgO + H_2O \rightarrow Zn(OH)_2 + Hg$$
$$\downarrow \quad \uparrow$$
$$ZnO + H_2O$$

and 2 faradays are released per mole of reagents. The depolarizer and cathode are mercuric oxide. Since the product of reaction, mercury, is a conductor, the internal resistance of the cell does not fall during use, and 'recuperation periods' are not necessary. There are many constructional variations.

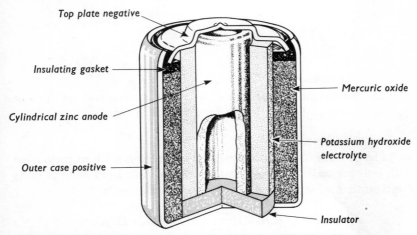

Top plate negative

Insulating gasket

Cylindrical zinc anode

Outer case positive

Mercuric oxide

Potassium hydroxide electrolyte

Insulator

Fig. 7.2. The original basic mercury cell which gives 1·34 V stable output.

The alkaline manganese cell

This is similar in construction to the mercury cell. It is used where a stable voltage and smallness are not so important. Its construction is shown in Fig. 7.3.

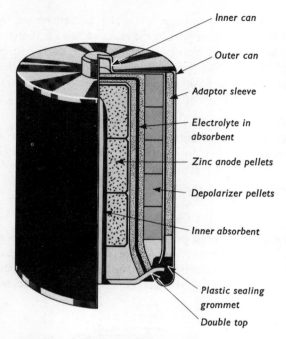

Inner can

Outer can

Adaptor sleeve

Electrolyte in absorbent

Zinc anode pellets

Depolarizer pellets

Inner absorbent

Plastic sealing grommet

Double top

Fig. 7.3. The alkaline manganese cell.

Secondary cells

The two important 'rechargeable' cells now in common use are the lead–acid type, and the nickel–cadmium alkaline type.

Lead–acid

In its basic form, the positive terminal is a perforated lead plate filled with lead peroxide, and the negative plate is lead. Both hang in fairly concentrated sulphuric acid. The reversible reaction is

$$\mathrm{Pb + PbO_2 + 2H_2SO_4 \underset{charge}{\overset{discharge}{\rightleftharpoons}} 2PbSO_4 + 2H_2O.}$$

During discharge the electrolyte loses sulphuric acid, and its density falls. Thus a hydrometer can be used to check the charge of a lead–acid cell. The electrolyte density of a fully charged cell is 1250 kg m^{-3}, and of a fully discharged one, 1100 kg m^{-3}.

During repeated charge and discharge, solid reaction products collect below the plates. When the level of these products reaches the suspended plates, the cell fails, so it is quite easy to predetermine roughly the age at which a lead–acid cell will fail.

The fully-charged e.m.f. of the lead cell accumulator is 2·05 V, and it falls only slightly during use.

Nickel–cadmium

The nickel–cadmium accumulator is lighter and more robust than the lead–acid. Its e.m.f., which falls during discharge, is 1·2 V. Both plates are made of perforated steel; the negative plate is filled with cadmium

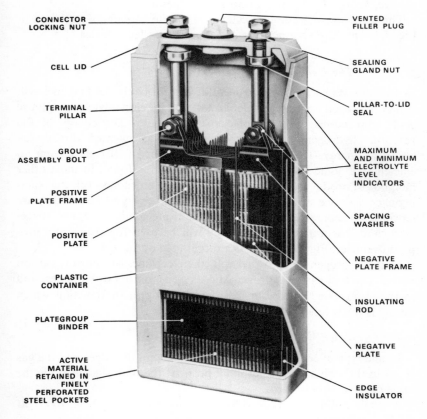

Fig. 7.4. A nickel–cadmium cell. Cutaway section of Alcad plastic-cased cell.

oxide and iron oxide, and the positive plate with nickel hydrate and graphite. The charge and discharge reaction is

$$2Ni(OH)_3 + Cd \underset{\text{charge}}{\overset{\text{discharge}}{\rightleftharpoons}} 2Ni(OH)_2 + Cd(OH)_2.$$

In a fully-charged cell the nickel hydrate is highly oxidized and the negative material is reduced to pure cadmium. On discharge the nickel hydrate is reduced to a lower degree of oxidation, and the cadmium in the negative plate is oxidized. Thus the reaction may be regarded as the transfer of $(OH)^-$ ions from one plate to the other, and the density of the electrolyte (21 per cent potassium hydroxide solution) does not change, being 1200 kg m^{-3} at normal temperatures.

The latest nickel–cadmium cells have plastic containers as shown in Fig. 7.4. They are very popular in battery-propelled vehicles.

The fuel cell

In 1839 Sir William Grove made what was probably the first fuel cell. It had hydrogen and oxygen electrodes, and an electrolyte of sulphuric acid; connections were via blackened platinum. The energy of the reaction (hydrogen + oxygen gives water) was released directly as electrical energy. Recent research has produced fuel cells having an efficiency of some 80 per cent, far higher than is found in most other methods of producing electricity. Whereas in an ordinary cell the reacting chemicals are stored in the reaction chamber, in a fuel cell they are fed in from separate containers. Fuel cells are used in manned spacecraft, and it is anticipated that they will be used for electrically propelled vehicles. Figure 7.5 shows a recent type of fuel cell.

In the hydrogen–oxygen fuel cell, the electrodes are porous carbon plates. The fuel gases (hydrogen and oxygen) ionize on the plates, and in the electrolyte (potassium hydroxide solution), the reaction which occurs is the reverse of the electrolysis of water:

$$2H_2 + O_2 = 2H_2O.$$

In one type of cell, two plates are sealed round the edges and the gas is fed in through a hollow terminal post. At the opposite electrode the same construction is used.

A single hollow-electrode cell of average size will deliver 100 A at 0·6 V continuously.

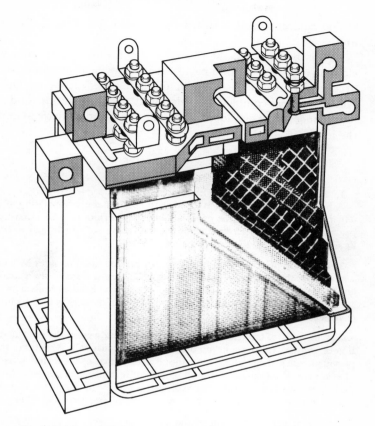

Fig. 7.5. Nine-plate hydrogen–oxygen cell.

Summary

Faraday's laws of electrolysis

(1) The mass of a substance liberated in electrolysis is proportional to the quantity of electricity passed, i.e. to the product of current and time.

(2) When the same quantity of electricity is passed through different electrolytes the masses of substances liberated are in the ratio of their equivalent weights (relative atomic mass ÷ charge on each ion).

The electrochemical equivalent of a substance is the mass liberated in electrolysis by 1 coulomb.

The quantity of electricity which liberates 1 mole of all singly-charged ions, $\frac{1}{2}$ mole of doubly-charged ions, and so on, is approximately

96 500 coulombs and is called the *faraday*. The faraday is not an SI unit.

As has been stated, the theory of Arrhenius assumes that when an electrolyte enters solution part of it is dissociated into ions, the extent of dissociation increasing with dilution. The theory of Debye and Hückel assumes complete dissociation at all concentrations and accounts for the increase in conductivity on dilution by an increase in mobility of the ions.

Questions

1. State Faraday's Laws of Electrolysis.
A brass plate of total area 100 cm^2 is to be given a copper plating of thickness 0·01 mm. How long will the process take if the current is 1 A? (Density of copper = 8930 kg m^{-3}; electrochemical equivalent of hydrogen = $1·04 \times 10^{-8}$ kg C^{-1}; chemical equivalent of copper = 31·6.)

2. Calculate the volume of mixed gases liberated in 10 minutes by a current of 1 A passing through a water voltameter. The temperature of the gases is 293 K and their pressure is 765 mm of mercury. The electrochemical equivalent of hydrogen is $1·04 \times 10^{-8}$ kg C^{-1}, the density of hydrogen at s.t.p. is 0·09 kg m^{-3} and of oxygen 1·43 kg m^{-3}. (B)

3. One form of electricity quantity meter depends upon the electrolytic liberation of mercury from a solution of a divalent mercury salt. A meter of this type is provided with a shunt whose resistance is $\frac{1}{9}$th of that of the meter. What quantity of electricity has been used when 1·39 cm^3 of mercury is deposited? Assume that the electrochemical equivalent of hydrogen is $1·045 \times 10^{-8}$ kg C^{-1}, that the atomic weight of mercury is 200·6 and that the density of mercury is 13 570 kg m^{-3}. (N)

4. Give a succinct account of the elementary theory of electrolysis and show that it is consistent with Faraday's laws of electrolysis.
What current will liberate 100 cm^3 of hydrogen at 293 K and 78 cm of mercury pressure in 5 minutes? (E.C.E. of hydrogen = $1·044 \times 10^{-8}$ kg C^{-1}; density of hydrogen = 0·09 kg m^{-3} at N.T.P.) (L)

5. A sheet of copper is suspended by copper threads midway between two copper electrodes in a bath of copper sulphate solution and these two electrodes are connected to the poles of a cell. Describe and explain what will happen to all three sheets of copper in the course of a day. (D)

6. In calibrating an ammeter by means of a copper voltameter it was found that the gain plate increased by 5·9 g h^{-1} when the average reading under test was 4·5 A. What was the percentage error of the ammeter? (Chemical equivalent of copper = 31·65; a faraday = 96 540 C.) (O S)

7. Given that the faraday is $9·64 \times 10^4$ C mol^{-1} and the atomic charge $1·59 \times 10^{-19}$ C, calculate the number of atoms in a mole, and the mass in g of the hydrogen atom. (B)

8. Give an account of an experimental method for determining the electrical conductivity of an electrolyte. What do you understand by *equivalent conductivity*? How does it vary with the concentration of the electrolyte? (N)

9. Write an account of electrolytic and metallic conduction pointing out their resemblances and differences.

Discuss the dissociation theory of electrolytic conduction, paying particular attention to the experimental facts which support it. (O & C)

10. What are Faraday's laws of electrolysis? Describe how you would verify them.

What evidence do these laws provide for the belief that there is a fundamental unit of electric charge, and how could the results of the experiments you describe be used to estimate its magnitude? (C S)

11. Explain the phenomenon of polarization in a simple cell and describe some experiments which support your statement.

Two accumulators, each of e.m.f. 2 V and of negligible internal resistance, are connected in series with a voltameter containing dilute sulphuric acid. The current which flows is $1 \cdot 012$ A. A third 2 V accumulator is then added in series and the current rises to $1 \cdot 812$ A. Calculate the resistance of the dilute acid between the electrodes. (D)

12. A water voltameter and a copper voltameter of resistances $2 \, \Omega$ and $1 \, \Omega$ respectively are connected in parallel, and a battery of e.m.f. 6 V and internal resistance $0 \cdot 5 \, \Omega$ is used to send current through them. Find the current in each of the voltameters, assuming that the e.m.f. of polarization of the water voltameter is $1 \cdot 8$ V. (N)

13. Two platinum plates are immersed in acidulated water and connected to the terminals of a voltmeter. They are also connected through a key and an ammeter to a battery of e.m.f. about 4 V. The key is held down for a few minutes and then raised. Describe and account for the indications of the voltmeter and of the ammeter from the time when the key is first closed until a few minutes after it had been opened. (D)

14. A 12 V car battery of total resistance $0 \cdot 2 \, \Omega$ is to be charged from 200 V d.c. mains, using a current of about 5 A. How many 200 W 200 V lamps will be required to form a suitable resistance? Draw a diagram of the circuit. How much will it cost to charge the battery for 10 hours, if the price of electrical energy is 1p per kilowatt-hour? What percentage of the energy is dissipated in heat? (C)

15. Describe briefly the lead secondary battery and explain the chemical reactions which take place in it during charging and discharging.

Such a battery is charged at different voltages (E) and the current (I) which flows through it is measured. The following observations are made:

E (volts)	$2 \cdot 3$	$2 \cdot 6$	$3 \cdot 0$
I (amperes)	$0 \cdot 4$	$1 \cdot 6$	$3 \cdot 2$

How do you reconcile these facts with Ohm's law? If this battery is charged at $3 \cdot 5$ volts, what is the maximum efficiency of the process (i.e. ratio of energy of chemical transformation to total energy supplied)?

16. Describe the changes which take place in the lead storage cell during charging and discharging.

Why is it bad for the cell to charge or discharge at too high a rate or for too long a time? Consider each case separately. Contrast the value of a hydrometer and a voltmeter in determining the extent to which a cell is discharged.

17. Describe the nickel–cadmium accumulator. What are the charge and discharge reactions? What advantages has it, compared to the lead–acid cell?

8   Electrostatics

So far in this book we have discussed phenomena connected with the electric current, i.e. electric charges in motion. We shall now consider electric charges at rest, a study known as electrostatics.

Charges by rubbing

Electrostatics began long before the invention of the electric cell by Volta and the consequent investigation of the electric current at the beginning of the nineteenth century. Indeed, the word 'electricity' was coined by William Gilbert in the reign of Queen Elizabeth I from the Greek word for amber, electron.

The ancient Greeks discovered that amber when rubbed with silk acquires the property of attracting light objects such as pieces of chaff. Gilbert discovered that other substances exhibit the same effect, and that the magnitude of the effect is roughly proportional to the area of the surface rubbed. He was thus led to the idea of a charge of electricity.

About 1745, Du Fay discovered that there are two kinds of electricity. Two ebonite rods when rubbed with fur exert a force of repulsion on each other. Two glass rods rubbed with silk also repel one another. However, an ebonite rod which has been rubbed with fur attacts a glass rod which has been rubbed with silk.

Any substance rubbed with a different substance acquires a charge of electricity, and will be found either to repel charged ebonite and to attract charged glass, or vice versa. Since the two kinds of electricity can neutralize each other's effects, one is called positive and the other negative, the choice as to which is positive being purely arbitrary. Glass rubbed with silk is said to have a positive charge and ebonite rubbed with fur a negative charge. Cellulose acetate $(+)$ and polythene $(-)$ – both rubbed on cloth – are superior modern alternatives.

The law of force between charges may be stated thus: *like charges repel, unlike charges attract.*

Ebonite and glass become charged only in the area in which they have been rubbed but a metal such as brass, on being rubbed in one part, is found to be charged all over. This is explained by the fact that ebonite

LUIGI GALVANI   ALESSANDRO VOLTA

LUIGI GALVANI (1737–1798), *professor of anatomy at Bologna, during the course of experiments on the nerves of the frog, discovered that a freshly dissected frog's leg twitched when placed near to an electrical machine which was generating sparks. Subsequently he found that the leg twitched when touched by two dissimilar metals such as copper and zinc. He believed that the source of the electricity was in the frog and was thus led into controversy with Volta who held, and later proved, that this was not 'animal electricity', but of the same nature as that generated by rubbing. Nevertheless, Galvani's single-minded investigation of a phenomenon of wholly unforeseen importance amply earned the fame of commemoration in the name 'galvanometer'. During eleven years of arduous experiment, mainly with frogs, he observed the effects (as is now known) of electromagnetic induction and of electromagnetic waves.*

ALESSANDRO VOLTA (1745–1827), *professor of physics at Pavia, invented an electroscope consisting of two straws – the forerunner of the gold leaf electroscope – and also the electrophorus. He investigated the effect discovered by Galvani and showed first that a frog's leg twitched when a discharge from a capacitor was passed through the nerve. He then demonstrated that different pairs of metals, when touching the frog's leg, gave effects of different magnitudes and he arranged the metals in an electromotive series. Finally, in a brilliant experiment, considering the smallness of the potential difference involved, he deflected an electroscope, by combining it with a capacitor, from the contact of a single pair of metals, thus dispensing with the frog. His invention of the 'voltaic' cell opened a new era in the history of electricity. He travelled widely, was a friend of Voltaire, Lavoisier and Priestley, and was a favourite of Napoleon.*

and glass are insulators and do not allow electricity to pass through them, whereas metals are conductors. If the charge acquired by a metal when rubbed is to be retained, the metal must be held with an insulating handle; otherwise the charge will escape through the metal and the human body to the earth. Charges can pass through moisture on the surface of a substance, and in all experiments in this branch of electricity apparatus must be kept dry.

The electroscope

The simplest method of detecting the presence of a charge on a body is by the attraction of a light object. One of the earliest instruments, the pith-ball electroscope, consisted of a small ball of pith, the white substance of very low density found under the bark of an elder tree. If the pith ball is suspended by a silk thread which, when dry, is an excellent insulator, it can be given a permanent charge by touching it with a rubbed rod. It will then serve to detect the sign of a charge; if, for example, its charge is negative, it will be attracted by a positive charge and repelled by a negative.

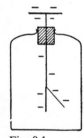

Fig. 8.1

A more convenient instrument is the gold-leaf electroscope. This consists of a leaf of thin metal foil suspended close to a vertical metal rod or plate to which it is attached along its upper edge. When the rod and leaf are charged the repulsion between like charges causes the leaf to diverge as in Fig. 8.1. The leaf is protected from draughts by enclosure in a metal case with glass windows. The rod is usually provided with a metal disk at its upper end and must be insulated from the case with a good insulator such as polythene. Note that in Fig. 8.1 the plug is shaded, a convention which will be used for insulators, while conductors will be left unshaded.

The instrument may be charged negatively by flicking the disk with fur. Then on bringing near to the disk a negatively charged body the

leaf diverges farther. This can be explained by the repulsion of the negative charge on the disk down to the leaf and lower part of the rod. A
positively charged body causes the leaf to fall owing to the attraction of
negative charge up to the disk.

The electric field

A charged body exerts an influence in the space around it and is said to
create an *electric field*. If a second charged body is placed in the electric
field of the first charged body, it will tend to move – to be attracted or
repelled. This is a fundamental characteristic of an electric field, that it
exerts a force on a charge placed in it.

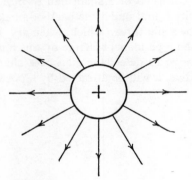

Fig. 8.2

The electric field can be investigated by scattering certain crystals
(for example, gypsum) on an insulating plate placed over the charged
body. The crystals set themselves in lines called *electric field lines*
(Fig. 8.2). The positive direction of the lines is indicated by arrow-heads
and is taken as the direction in which a small positive charge would tend
to move.

Electric potential

We have seen that a current of electricity, which consists simply of
moving charges, can flow only when there is a difference of potential.
Thus in an electric field there must be a potential gradient. If the
potential gradient is constant, the field is said to be uniform, and the
steeper the potential gradient, the stronger the field.

The meaning of electric potential can be illustrated by means of an
analogy. A body when raised above the surface of the earth is said to
acquire potential energy because it can do work in falling; if free to

move in a gravitational field it will fall to the position in which its potential energy is zero. Similarly, a charged body in an electric field has potential energy, and it will tend to move to those parts of the field where its potential energy is smaller. When a positive charge is repelled by another positively charged body and moves away, its potential energy decreases. Its potential energy will be zero when it is completely away from the influence of the charged body, i.e. at infinity.

The electric potential at a point, in volts, is defined as the work done, in joules, in bringing up a charge of 1 coulomb from infinity to the point.

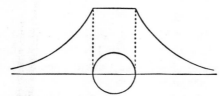

Fig. 8.3

Potential due to a charged spherical conductor

For a charged spherical conductor the potential varies with the distance from the centre in the manner represented in Fig. 8.3. The height of the graph above a point on the diameter produced represents the potential at that point. Note that the potential is uniform throughout the sphere and falls off in the space surrounding it.

Capacitance

When a charge is given to a conductor its potential rises, just as when heat is given to a body the temperature rises. To produce the same rise in potential it is necessary to supply a larger charge to a large body than to a small body. We say that the large body has a greater electrical *capacitance* or capacity than the small body.

The *capacitance of a conductor* is *its charge divided by its potential,*

i.e. $$C = \frac{Q}{V},$$

where C = capacitance, Q = charge, V = potential

If 1 coulomb raises the potential of a conductor by 1 volt, the capacitance of the conductor is 1 *farad*.

The potential of the earth

The earth is so large a conductor that any small charges it may have on its surface will not raise its potential appreciably, and we can assume that its potential is constant. Its potential is taken as zero, that is, the same as a region remote from all electrical charges. Similarly, mean sea-level is taken as the zero for the heights of mountains and the depths of seas.

The potential of a conductor may be regarded as positive if, when connected to the earth, positive electricity flows from the body to the earth, or alternatively, negative electricity flows from the earth to the body. The conductor has a negative potential if the flow of electricity is in the opposite direction.

Electrostatic induction

When an uncharged conductor is placed in an electric field there is a rapid flow of charge in it until its potential is the same throughout; positive and negative *induced* charges appear at its ends. The phenomenon is known as electrostatic induction.

Figure 8.4 illustrates the charges induced in a cylindrical conductor by a positively charged sphere; it shows also the potential distribution. The height of the graph above each point on the straight line through the centres of the sphere and cylinder represents the potential at that point. Note that the potential throughout the cylindrical conductor is the same and that therefore the field has disappeared inside it. This change in the field is, of course, due to the induced charges; the negative induced charge lowers the original potential at one end, and the positive induced charge raises it at the other.

The existence and sign of the induced charges may easily be verified by means of a proof plane, which consists of a metal disk on an insulating handle. The proof plane is laid on one end of the cylindrical conductor and when removed carries away with it a sample of the induced charge which can be tested with a gold-leaf electroscope.

The induced charges are temporary, and disappear when the field is removed. If, however, the cylindrical conductor in Fig. 8.4 is touched to earth before the field is removed, it will be found to have a permanent

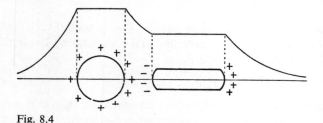

Fig. 8.4

negative charge. This can be explained by assuming that the induced positive charge was repelled to earth by the positive charge on the sphere, while the induced negative charge was held in position by attraction. Alternatively, we can say that the cylindrical conductor had a positive potential, caused by the presence of the positively charged sphere, and hence positive electricity flowed to earth until its potential was zero.

Electrostatic induction occurs only in conductors and semiconductors, since electricity cannot flow in insulators. A good example of induction in a semiconductor is the insulated gate field effect transistor.

Theories of electricity

In electrification by friction equal and opposite charges appear on the rubber and the rubbed. Thus if a polythene rod is rubbed with a piece of cloth mounted on an insulating handle, and the two are presented together to a charged gold-leaf electroscope, the leaf does not move, since they neutralize each other's effects. When the polythene and cloth are brought up separately, however, the polythene is seen to be negatively charged and the cloth positively.

This fact, together with the phenomenon of electrostatic induction, suggests that all uncharged bodies contain equal quantities of positive and negative electricity.

In the eighteenth century it was suggested that there were two weightless, electric fluids, and that an electric current consisted of a flow of each in opposite directions. Benjamin Franklin proposed, for the sake of simplicity, that an electric current should be regarded as a flow of positive electricity only, a convention still retained. It is now known, however, that in a metal a current consists of a flow of negative electrons, whereas in a liquid electrolyte it consists of a flow of both positively and negatively charged particles in opposite directions.

All matter is now believed to be composed of electrical charges. An atom consists of a small, massive nucleus of positive electricity with a number of negative charges, called electrons, surrounding it. An atom of copper has twenty-nine planetary electrons and a nucleus with an equivalent positive charge. The outermost electron of copper can easily break away from the atom. In all metals one or two of the outermost electrons can easily become free in this way. These free electrons form a kind of electron gas which moves slowly under the influence of an electric field and constitutes a current.

Metals viewed under the microscope are seen to be mosaics of microcrystals; in each of these tiny crystals the atoms are fixed by their mutual attractions in a simple, openwork arrangement known as a lattice, through which the free electrons move.

The process of induction, illustrated in Fig. 8.4, is therefore explained as a movement of electrons from right to left, giving the left-hand end of the cylindrical conductor an excess of electrons, and hence a negative charge, and the right-hand end a deficiency of electrons, and hence a positive charge.

When the cylindrical conductor is connected to earth, instead of positive electricity passing from the conductor to the earth, negative electrons pass up from the earth to the conductor.

The atoms of insulators do not readily release electrons and hence contain no free electrons to constitute a current. But in the intimate contact with another substance necessitated by rubbing, they may either gain or lose electrons, and hence become charged in the region where they have been rubbed.

Charging a gold-leaf electroscope by induction

Bring a rubbed ebonite rod or polythene strip near to the disk of a golf-leaf electroscope. Charges are induced in the electroscope as in Fig. 8.5 (*a*), and the leaf rises; electrons have been repelled from the disk to the lower end of the electroscope. Without moving the ebonite rod, earth the electroscope by momentarily touching the disk; electrons run from the electroscope to the earth and the leaf falls (Fig. 8.5 (*b*)). Remove the ebonite rod and the leaf rises (Fig. 8.5 (*c*)); electrons have run from the leaf to the disk, leaving the leaf positively charged and reducing the positive charge on the disk.

The electroscope measures potential rather than charge

In Fig. 8.5 (*a*) the electroscope has a negative potential owing to the presence of the negatively charged ebonite rod, but its resultant charge

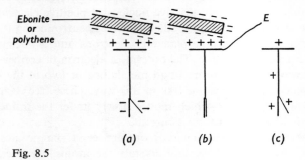

Fig. 8.5

is zero; the leaf diverges. In Fig. 8.5 (*b*) the electroscope is at zero potential but it has a resultant positive charge; the leaf does not diverge.
Thus the divergence of the leaf indicates the potential rather than the
charge of the electroscope.

If a positively charged body is brought near to the electroscope of
Fig. 8.5 (*c*), the positive potential of the electroscope increases and
hence the leaf diverges farther. If a negatively charged body is brought
near, the positive potential of the electroscope is reduced and the leaf
falls. Thus the electroscope may be regarded as an electrostatic voltmeter.

The charge resides on the surface

The charge on a conductor resides on its outside surface owing to self
repulsion. This can be demonstrated by charging an insulated hollow
vessel, such as a deep metal can, and testing samples of the charge on
various parts of its inside and outside surfaces with a gold-leaf electroscope by means of a proof plane (consisting of a metal disk on an
insulating handle). The proof plane will be found to carry away no
charge when touched deep inside the can. Faraday used a charged,
conducting, linen butterfly net (Fig. 8.6); on turning the net inside out
with an insulating silk thread, he found that the charge changed surface
so that it resided once more on the outside.

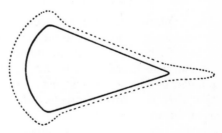

Fig. 8.6   Fig. 8.7

The distribution of charge

The charge on an isolated, spherical conductor is uniformly distributed
on its surface, but in the case of a conductor which has not a constant
curvature the charge is more dense on the more curved parts. This may
be demonstrated by taking samples of the charge from various parts of
the surface with a proof plane and testing them with a gold-leaf electroscope. In Fig. 8.7 the variation in the distance of the dotted line from

the surface of the conductor indicates the variation in the surface density of the charge.

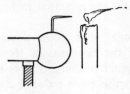

Fig. 8.8

The action of points

At the tip of a pointed conductor the curvature is very great, and hence there is a very high surface density of charge. This may be sufficiently great to charge the air and repel it in the form of an electric wind which can be demonstrated by its effect in blowing a candle flame (Fig. 8.8).

The air is in fact 'ionized'; the molecules are split into positively and negatively charged 'ions'. A positively charged point attracts the negative ions which tend to discharge it, and repels the positive ions to form the electric wind.

Ionization is often accompanied by a blue or violet glow known as a brush discharge which can be seen in the dark.

Lightning

The similarity between an electric spark and a lightning flash was noticed as soon as electrostatic generating machines were invented.

In 1752 Benjamin Franklin demonstrated, by utilizing the action of points, that thunderclouds are charged with electricity. Suppose a pointed conductor is held near to a negatively charged ebonite rod (Fig. 8.9). A positive charge is induced on the point and streams away as an electric wind which tends to neutralize the charge on the ebonite, while a negative charge is induced on the lower end of the pointed conductor.

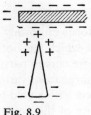

Fig. 8.9

Franklin attached a pointed conductor to a kite which he flew in a thunderstorm, and when the string became wet and a conductor, he was able to draw a spark from a key attached to the lower end of the string. The psychological effect of this famous experiment was much greater than its theoretical importance; it gave to a capricious and terrifying phenomenon a material explanation and related it to scientific law. Moreover, Franklin showed how the striking of buildings by a lightning discharge, previously ascribed to divine wrath, could be circumvented. A pointed conductor, efficiently earthed and attached to the highest point of a building, gives off an electric wind tending to neutralize a charged cloud above it; if, despite this action, the discharge still occurs, the conductor conveys it safely to earth.

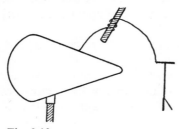

Fig. 8.10

Use of the electroscope to show that the potential is constant over the surface of a charged conductor

Charge a conductor by induction with an ebonite rod. Move a wire, wrapped round an insulating handle, and connected to an electroscope (see Fig. 8.10), over the surface of the conductor. The divergence of the leaf of the electroscope is constant, showing that the potential over the surface of the conductor is constant.

In this experiment the conductor and the electroscope are permanently connected. In the experiment described on p. 169 the proof plane was disconnected from the conductor, and when taken away from the influence of the latter it acquired a potential determined by its capacitance and the magnitude of the sample of charge it carried away; it was this potential that the electroscope measured.

Faraday's ice-pail experiment

Faraday performed famous experiments to show that the total induced charge is always equal to the inducing charge. He wrote that 'their value consists in their power to give a very precise and decided idea to the

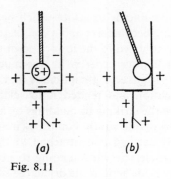

(a)   (b)

Fig. 8.11

mind respecting certain principles of inductive electrical action, which I find are by many accepted with a degree of doubt or obscurity that takes away much of their importance'.

He used an ice-pail, replaced in Fig. 8.11 by a deep can, and the following is one of the experiments. Place the can on an electroscope and lower a positively charged sphere into the can. The leaf diverges. When the sphere is touched to the inside of the can, the divergence of the leaf does not change and the sphere, on removal, will be found to be totally discharged.

In Fig. 8.11 (a) five positive charges on the sphere are shown as inducing five negative charges on the inside of the can and these neutralize in (b). The five induced positive charges on the outside of the can remain when the sphere touches the inside and also when it is withdrawn.

Faraday imagined lines of electric force to emerge from a charged body as shown in Fig. 8.2 on p. 164. The can must be deep so that all the lines of electric force from the charged sphere terminate on the can. The experiments show that lines of electric force which start on a positive charge always end on an *equal* negative charge. Faraday made this very clear by speaking of 'tubes of force', each starting on unit quantity of positive charge and ending on unit quantity of negative charge.

The capacitor

When an insulated metal plate A (Fig. 8.12) is given a positive charge and an earthed plate B is brought near to it, the positive potential of A is reduced, owing to the effect of the negative charge induced on B. If B is brought nearer, the potential of A is lowered still farther. This can be demonstrated by connecting A to a gold-leaf electroscope which measures its potential. The leaf of the electroscope falls as B is brought nearer to A.

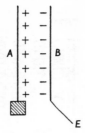

Fig. 8.12

The insulated plate's charge, Q, is unchanged, but its potential, V, is reduced by the presence of the earthed plate. Thus its capacitance, $C = Q/V$, has been increased. The arrangement of the two plates is called a capacitor, and is a convenient means of storing a comparatively large charge without an excessively high potential. It is an essential component of all electronic circuits.

Similarly, in Fig. 8.4 the potential of the sphere is lowered by the presence of the cylinder and hence its capacitance is increased. The capacitance of all bodies is influenced by the presence of nearby conductors.

Dielectrics

If a slab of an insulator, such as polythene, is placed between the plates of a charged capacitor, the potential difference between the plates falls and thus the capacitance increases. This can be demonstrated experimentally by connecting the positively charged plate of the capacitor to a gold-leaf electroscope. When the slab of polythene is inserted the leaf of the electroscope falls a certain distance, and rises again when the slab is withdrawn.

The ratio of the capacitance of the capacitor with the insulator (called a dielectric) filling the space between the plates, to its capacitance with a vacuum between the plates is called the *relative permittivity* or the dielectric constant of the dielectric, and is denoted by ϵ_r.

Some values of ϵ_r are:

Air	1·0006	Sulphur	3·6–4·3
Dry paper	2–2·5	Mica	5·7–7
Paraffin wax	2–2·3	Glass	5–10
Ebonite	2·7–2·9	Pure water	81

The fall in potential difference between the plates is caused by 'polarization' of the dielectric. Under the action of the electric field the

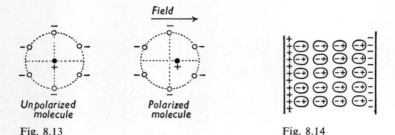

Fig. 8.13   Fig. 8.14

molecules become distorted, the electrons being attracted towards the positive plane and the positive nuclei towards the negative plate, as shown diagrammatically in Fig. 8.13. The molecules may be regarded then as having positively and negatively charged ends, that is, as electric dipoles.

Positive and negative ends of adjacent molecules (Fig. 8.14) neutralize each other's effects, but the surface of the dielectric against the positive plate has a negative charge and this lowers the potential of the plate. Similarly, the positive charge at the other surface of the dielectric makes the potential of the negative plate less negative.

The molecules of certain substances with a high relative permittivity, such as water, are permanent electric dipoles; that is to say, the 'centres of gravity' of the positive and negative charges do not coincide even when there is no electric field. Under the action of an electric field they align themselves rather like magnetic compasses in a magnetic field. This effect is superimposed on the normal effect and accounts for the high dielectric constant.

The analogy with magnetism is even more striking in the case of a substance known as carnuba wax. This becomes permanently polarized when melted and allowed to solidify in a strong electric field. The molecules are 'frozen' into alignment. If the wax so treated is cut into pieces, each piece will set itself in an electric field with its positive face in the direction of the field.

Residual effect

When the plates of a charged capacitor are connected, a spark passes immediately before contact is made and the plates are discharged. With a dielectric such as glass, several subsequent but much smaller sparks can be obtained without recharging. This is known as the residual effect, and may be explained by the dielectric retaining some of its polarization for some time. As the polarization slowly disappears, charges on the plates are released from the attraction of the dielectric and enable a small spark to be obtained.

An illuminating experiment can be performed with a capacitor having a dielectric such as glass and removable plates mounted on insulating supports. After charging, say with a battery, the plates are removed from the glass and discharged. On replacing the plates against the glass a discharge can be obtained from the capacitor. The charge must have remained on the glass when the plates were removed.

Practical forms of capacitors

To possess a high capacitance a capacitor must have plates of a large area as close together as possible and a dielectric with a high relative permittivity. A common form of radio capacitor consists of interleaved plates of tinfoil separated by thin sheets of mica or paraffin waxed paper (see Fig. 8.15). Sometimes a thin film of silver is deposited on both sides of thin strips of mica or ceramic base material.

Fig. 8.15

Variable capacitors usually take the form of a set of metal vanes which can move between a set of fixed metal vanes, air being the dielectric (see Fig. 8.16). The moving plates of early radio capacitors were semicircular, in which case the capacitance is proportional to the angle of overlap. The ends of the tuning scale of the radio receivers were cramped because the wave-length is proportional to the square root of the capacitance. The shape of the moving vanes was therefore altered so that the capacitance increased as the square of the angle of overlap. These capacitors are called square-law types.

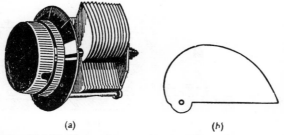

(a)   (b)

Fig. 8.16 (b) Shape of square-law capacitor plate.

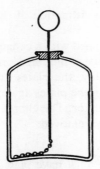

Fig. 8.17

The Leyden jar consists of a glass jar as the dielectric, with inner and outer coatings of tinfoil to act as the plates (see Fig. 8.17). The inner insulated plate is connected by a chain to an insulated knob for convenience of charge and discharge.

The electrophorus

The earliest machines for generating electric charges were friction machines. Otto von Guericke cast a globe of sulphur and caused it to rotate while rubbing it with a dry hand. Later the bare hand was covered with a leather glove, and then replaced by a pad. Modern machines generate charges by induction instead of by friction.

The electrophorus, invented by Volta, consists of a disk of some insulator such as ebonite which, on rubbing with fur or flannel, acquires a negative charge (see Fig. 8.18). A brass disk with an insulating handle, called a cover, is placed on the charged ebonite and touched to earth. It becomes charged positively by induction. On raising the cover away from the influence of the negative charge it acquires a considerable

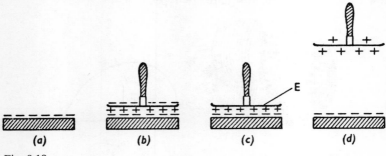

Fig. 8.18

positive potential and will provide a spark. The energy of the spark is derived from the mechanical work done in raising the cover against the attraction of the unlike charges. Numerous charges may be obtained by repeating the process without any further rubbing of the ebonite. The latter is usually mounted on a metal disk called a sole; a positive charge is induced on the sole which tends to prevent the charge on the ebonite from leaking away.

Van de Graaff's generator

Figure 8.19 represents an electrostatic generator, built in America under the direction of Van de Graaff, which depends for its working on the action of points.

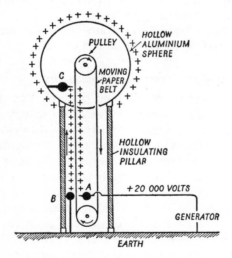

Fig. 8.19. The Van de Graff generator.

A pointed conductor A is attached to the positive terminal of a generator producing d.c. at 20 kV. A positively charged electric wind flows from the point on to a moving paper belt which collects the charge and carries it up into an aluminium sphere, 5 m in diameter, supported on hollow insulating columns 8 m high and 2 m in diameter. The earthed conductor B becomes negatively charged by induction and helps to attract the positively charged wind on to the paper belt.

The positive charge on the paper belt is carried up inside the sphere and induces a negative charge on the point of C and a positive charge on the outside of the sphere. A negatively charged wind flows from the point of C and neutralizes the positive charge on the paper belt. Thus the effect of C is to transfer the charge from the paper belt to the outside of the sphere.

By using two spheres and charging one positively and the other negatively, it is possible to obtain a p.d. of about 10 MV. This p.d. is used for producing a beam of electrons, of very high speed, for nuclear research. The apparatus is contained in an insulating tube between the spheres and the experimenters take up a position inside one of the spheres in which there is a room. Here they are perfectly safe since, although the sphere is charged to a potential of several million volts, all its charge resides on the outside.

The work done in charging the spheres is done by the motor driving the belt; there is a strong repulsion between the positive charge on the belt and the positive charge on the sphere.

The whole apparatus is housed in an airship hangar and can be moved outside the hangar on trucks which run on rails.

The identity of static and current electricity

The static electricity produced by rubbing or by a Van de Graaff generator is identical in nature with the current electricity produced by an electric cell or dynamo. The apparent difference in their effects is due to the fact that static electricity is usually at a very high potential but small in quantity, whereas current electricity is usually generated at a low potential difference but in comparatively large quantity. The e.m.f. of a simple cell is 1 V, whereas the p.d. required to produce a spark 1 cm long in air is of the order of 30 kV. The shock from a small Van de Graaff machine is not fatal because the quantity of electricity it generates is small; a shock from a dynamo or cells at a p.d. higher than 250 V is extremely dangerous, since the quantity of electricity passing depends

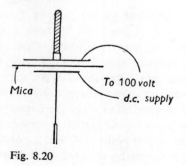

Mica   To 100 volt
          d.c. supply

Fig. 8.20

almost solely on the resistance of the human body; also the supply is continuous.

A battery or direct current generator giving a p.d. of a few volts will cause a divergence of the leaf of an electroscope if the latter is combined with a capacitor (see Fig. 8.20). The p.d. is applied between the top of the electroscope and a metal disk with an insulating handle, the two being separated by a thin sheet of mica or waxed paper. When the leads to the generator are disconnected and the disk is then raised, the leaf of .the electroscope will diverge. The disk is oppositely charged to the electroscope, and when it is removed the potential of the electroscope rises considerably.

Quantitative electrostatics

Coulomb's experimental proof of the law of force between point charges

The law governing the force between two point charges was discovered experimentally by Coulomb in 1784.

Coulomb charged two gilded pith-balls with like charges of electricity and found how the force of repulsion between them varied with their distance apart. One pith-ball was attached to the end of a light insulating rod suspended from a fine silver wire (Fig. 8.21) and the other was fixed. The repulsion between them was estimated from the twist of the wire. Coulomb discovered that the force is inversely proportional to the square of the distance between the balls, i.e.

$$F \propto \frac{1}{d^2}.$$

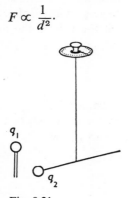

Fig. 8.21

If the charge on a pith-ball is shared with an exactly similar ball the charge on the original ball is halved. If this is again shared with a third similar ball the charge becomes one-quarter of the original charge, and so on. By using different submultiples of charge on the two balls in the torsion balance, at a fixed distance apart, Coulomb showed that the force between the charges is proportional to the product of their magnitudes.

The balls in Coulomb's experiment must be small compared with their distance apart. Otherwise, owing to repulsion, the charges are not uniformly distributed on the spheres, and it is unjustifiable to take the distance apart of the centres of the spheres as the mean distance apart of the charges. We can avoid this difficulty by the concept of point charges.

We express Coulomb's law by

$$F = \frac{1}{4\pi\epsilon}\frac{Q_1 Q_2}{d^2}.$$

The constant ϵ depends on the material surrounding the charges, and is called *permittivity*. We shall see later that it is advantageous to have the additional constant 4π in any system having 'spherical symmetry', i.e. any system in which effects are the same anywhere on the surface of a sphere. Likewise, in any system having 'cylindrical symmetry', it is an advantage to have a 2π in the formulae. Any system of consistent formulae employing this convention is called a 'rationalized system'. For a vacuum, $\epsilon = 8 \cdot 85 \times 10^{-12}$ F m^{-1} and is usually written ϵ_0.

Electric field, E

Any region in which an electrostatic force acts on any isolated charge is called an electric field. The magnitude of the field is the force which would act on unit charge, so

$$E = \frac{F}{Q}$$

and E could be measured in newtons per coulomb.

We can find the field at a distance d from a charge Q_1 by substituting 1 for Q_2 in the formula expressing Coulomb's Law:

$$E = \frac{F}{Q_2} = \frac{1}{4\pi\epsilon}\frac{Q}{d^2}.$$

Electric potential, V

The potential difference between two points has been defined as the work needed to move unit charge from one to the other. The potential *at* a point is defined as the work needed to bring unit charge to that point from an infinite distance. Applying this to the potential distance *d* from a point charge *Q* (see Fig. 8.22),

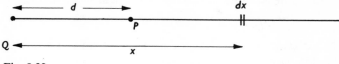

Fig. 8.22

we first find the work needed to bring the test (unit) charge a small distance *dx* closer, when it is already at a distance *x*. At this distance, the force on it is

$$F = QE = 1 \times \frac{1}{4\pi\epsilon} \frac{Q}{x^2}$$

so the work needed to bring it *dx* nearer is

$$dW = \frac{1}{4\pi\epsilon} \frac{Q\,dx}{x^2}.$$

The total work to bring the test charge from infinity to *P* (distance d) is therefore

$$W = \int_{\infty}^{d} \frac{1}{4\pi\epsilon} \frac{Q\,dx}{x^2}$$

$$= \frac{Q}{4\pi\epsilon d}.$$

Relation between *V* and *E*

Looking at the foregoing equations for field *E* and potential *V*, we see that *E*, already defined by the equation *F/Q*, is also numerically equal to *V/d*. In fact, it is usually more helpful to quote the value of *E* as so many volts per metre. Examining the units,

$$\frac{newton}{coulomb} = \frac{newton\text{-}metre}{coulomb\text{-}metre} = \frac{joule}{coulomb\text{-}metre} = \frac{volt}{metre}.$$

Charged spheres

The foregoing formulae for field and potential at a distance d from a point charge *Q* also apply if the charge is spread over the surface of a sphere of radius *r* ≤ d. This follows from the fact that we represent an electric field in diagrams by lines which start from a certain number of + charges, and eventually terminate on an equal number of −charges.

Whether the charge is concentrated at the centre of a sphere, or spread evenly over its surface, the arrangement at and outside the surface must be the same. Further, since all points on a conducting sphere must be at the same potential, there can be no field *inside* a closed conductor.

Electric flux

In a diagram showing electric field lines, lines which are closer together imply a stronger field. A certain number of lines originate from a certain amount of charge, and the concentration of lines through any particular area is called the electric flux:

$$\text{Electric flux} = E \times \text{area}$$

For example, we will look at the flux passing through the surface of a sphere of radius r, surrounding a charge Q. At the surface of the sphere, the electric field is given by

$$E = \frac{1}{4\pi\epsilon}\frac{Q}{r^2}$$

and the area of the sphere is $4\pi r^2$. Thus the total flux through the surface of the sphere of area A is

$$EA = \frac{1}{4\pi\epsilon}\frac{Q}{r^2}\,4\pi r^2 = \frac{Q}{\epsilon_0}.$$

Note that this is independent of the radius of the sphere. Now, as each line on a diagram represents a certain amount of flux, we can draw any other shape to enclose the charge Q, and the same number of lines, i.e. the same flux (Q/ϵ) must emanate from the charge.

Further, if more than one charge is in the space, the flux through its boundary is the algebraic sum of the separate fluxes, viz.

$$\frac{Q_1 + Q_2 + Q_3 + \cdots}{\epsilon}.$$

This result is known as *Gauss' Law*.

Finally, it follows that if we draw an enclosure of any shape in any region of electric field, if there is no charge to be found *inside* the enclosure,

$$\text{Total flux entering the space} = \text{total flux leaving it.}$$

This applies equally for lines of magnetic flux; and in this context there exists a very nice experimental proof of its validity. One end of an iron retort stand is inserted through one side of a cube framework, and six separate coils entirely surround the cube space. A coil carrying a.c.

current is placed around the iron bar, outside the cube. Magnetic flux produced by the external coil enters the hollow cube, concentrated in the iron bar. When any number of coils are connected in series, an induced voltage is found, *except* when all six coils are put in series. In this case only, all induced e.m.f.'s cancel out, showing that the total magnetic flux entering the cube along the iron bar = total magnetic flux leaving it through the remainder of the boundary.

Capacitance

As was explained on p. 165, capacitance is determined from the formula

$$C = \frac{Q}{V}.$$

Capacitance of an isolated sphere

Since the potential at the surface of a conducting sphere is the same as it would be at that region if all the charge were concentrated at the centre, we have

$$C = \frac{Q}{V} = \frac{Q}{(1/4\pi\epsilon)\,(Q/r)}$$

$$= 4\pi\epsilon r.$$

Capacitance of concentric spheres

If a charge of $+Q$ is applied to the inner sphere, radius a, there must be a charge of $-Q$ on the outer sphere, radius b.

The potential of the outer sphere will be

$$V_b = \frac{1}{4\pi\epsilon}\frac{-Q}{b}$$

and the potential of the inner sphere will be

$$V_a = \frac{1}{4\pi\epsilon}\frac{+Q}{a}.$$

Thus the potential difference will be

$$\frac{1}{4\pi\epsilon}\left(\frac{Q}{a} - \frac{Q}{b}\right)$$

and the capacitance, Q/V, is

$$Q \div \frac{1}{4\pi\epsilon}\left(\frac{Q}{a} - \frac{Q}{b}\right) = \frac{4\pi\epsilon}{(1/a)-(1/b)}$$

$$= 4\pi\epsilon\,\frac{ba}{b-a}.$$

Capacitance of parallel plates

If a charge Q lies on the upper plate, the area of the plates being A, the total flux from it to the lower plate is $EA = Q/\epsilon$. But E is also given by V/d, where V is the voltage across the plates and d is the separation.

So the capacitance

$$C = \frac{Q}{V} = \frac{EA\epsilon}{V} = \frac{V}{d}\frac{A\epsilon}{V} = \frac{\epsilon A}{d}.$$

As most capacitors in practical use employ parallel plates of one kind or another, this result is particularly important.

Guard-ring capacitor

The formula is not accurate if the field lines at the edges of the plate are curved. The error can be eliminated by means of an experimental device known as a guard ring (Fig. 8.23). This consists of a ring RR surrounding the circular insulated plate P of the capacitor and nearly touching it. RR is kept at the same potential as P. The earthed plate has an area equal to the combined areas of P and RR. It can be seen from Fig. 8.23 that the field between P and the earthed plate is uniform; the non-uniform field at the edges of R does not matter since R is not part of the capacitor.

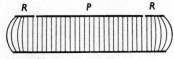

Fig. 8.23

It is of interest to note that the capacitance of a 'unit capacitor', consisting of parallel plates of area 1 m², separated by 1 m in a vacuum, is ϵ farads.

Relative permittivity; dielectrics

The formula $C = \epsilon A/d$ can be used to find the value of ϵ for the material between the capacitor plates. For very accurate results a guard-ring capacitor is used. A very simple method of measuring C is as follows (see Fig. 8.24):

R is a reed-switch. If the coil around it is connected to 50 Hz a.c., it will 'change over' 100 times a second, as the moving contact flicks regardless of the direction of magnetization. Alternatively a half-rectified supply may be used, in which case there will be 50 flicks per second.

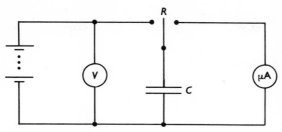

Fig. 8.24

As it takes only a millisecond or so for the capacitor, C, to charge to nearly the full supply voltage, V, C will acquire a charge $Q = CV$ every time the switch flicks; let us call that f times a second. Also, on each discharge, it loses nearly 100 per cent of its charge through the micro-ammeter. The latter is unable to follow the rapid short pulses, so it gives an average reading; the average current is

$$I = \frac{\mathrm{d}Q}{\mathrm{d}t} = Qf = CVf.$$

So, with I, V, and f known, C is calculated.

When a material other than air is inserted between the plates, higher values for ϵ are usually obtained. For example, if Perspex is used, the value $\epsilon_{\text{Perspex}}$ is found to be about 3·5 times the value ϵ_{air}. Careful experiments show that ϵ_{air} is only marginally grater than ϵ_{vacuum}, the latter being called 'the permittivity of free space', written ϵ_0. The ratio

$$\frac{\text{permittivity of material } x}{\text{permittivity of free space}}$$

is called 'relative permittivity', or 'dielectric constant', and is written ϵ_r. It has no units.

There is a relation between the permittivity of free space, and the magnetic permeability of free space,

$$\epsilon_0\mu_0 = \frac{1}{c^2}$$

where c = velocity of light $\simeq 3 \times 10^8$ m s^{-1}.

We shall not prove this relation but merely show that the units of $\epsilon_0\mu_0$ are the same as those of the reciprocal of the square of a velocity.

$$\text{Units of } \epsilon_0 = \frac{\text{farad}}{\text{metre}} = \frac{\text{coulomb/volt}}{\text{metre}} = \frac{\text{A s}}{\text{V m}}$$

$$\text{Units of } \mu_0 \;=\; \frac{\text{henry}}{\text{metre}} \text{ (p. 58)} \;=\; \frac{\text{V s}}{\text{A m}}$$

$$\text{Units of } \epsilon_0\mu_0 = \frac{\text{A s}}{\text{V m}} \times \frac{\text{V s}}{\text{A m}}$$

$$= \frac{\text{s}^2}{\text{m}^2}$$

$$= \text{units of } \frac{1}{(\text{velocity})^2}.$$

Since $\mu_0 = 4\pi \times 10^{-7}$,

$$\epsilon_0 = \frac{1}{\mu_0 c^2} = \frac{1}{4\pi \times 10^{-7} \times (3 \times 10^8)^2} = \frac{1}{36\pi} \times 10^{-9} \text{ F m}^{-1}.$$

Taking the exact value of the velocity of light, $2 \cdot 998 \times 10^8$ m s^{-1},

$$\epsilon_0 = 8 \cdot 854 \times 10^{-12} \text{ F m}^{-1}.$$

EXAMPLE

Calculate the force between two metal spheres, each of diameter 1 cm, with their centres 10 cm apart in air, when each is charged to a potential of 100 V.

We will assume that the charge on each sphere behaves as though concentrated at its centre. This is only an approximation because of the mutual repulsion of the charges on the two spheres.

$$\text{Capacitance of each sphere} = 4\pi a \epsilon$$
$$= 4\pi \times 0 \cdot 5 \times 10^{-2} \times \epsilon \text{ F}.$$
$$\therefore \text{ Charge on each sphere} = CV$$
$$= 4\pi \times 0 \cdot 5 \times 10^{-2} \times \epsilon \times 100$$
$$= 2\pi\epsilon \text{ C}$$
$$\text{Force between spheres} = \frac{Q_1 Q_2}{4\pi r^2 \epsilon_0}$$
$$= \frac{(2\pi\epsilon_0)^2}{4\pi(10 \times 10^{-2})^2 \epsilon_0}.$$

Taking $\epsilon_0 = 8 \cdot 86 \times 10^{-12}$ F m^{-1}.

Force between spheres $= 2 \cdot 8 \times 10^{-9}$ N.

Capacitors in parallel

The capacitors in Fig. 8.25 are said to be connected in parallel. It is clear that the three capacitors are equivalent to a single capacitor with

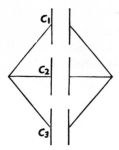

Fig. 8.25

plates of area equal to the sum of the areas of the plates of the separate capacitors, and hence that

$$C = C_1 + C_2 + C_3.$$

Capacitors in series or cascade

The capacitors in Fig. 8.26 are said to be connected in series (or cascade). When the extreme left-hand plate is given a charge Q, and the extreme right-hand plate is earthed, the outflow from each capacitor is equal to the inflow to the next, and the plates acquire charges Q and $-Q$ as shown. The total p.d. V across all the capacitors is equal to the sum of the p.d.'s across each.

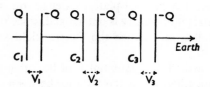

Fig. 8.26

Thus

$$V = V_1 + V_2 + V_3,$$

i.e.

$$\frac{Q}{C} = \frac{Q}{C_1} + \frac{Q}{C_2} + \frac{Q}{C_3}.$$

$$\therefore \frac{1}{C} = \frac{1}{C_1} + \frac{1}{C_2} + \frac{1}{C_3}.$$

Comparison of capacitances using a ballistic galvanometer

One of the capacitors C_1 is charged by a battery (Fig. 8.27) to a potential V, and then connected to a ballistic galvanometer whose throw, θ_1, is proportional to the charge, Q_1, of the capacitor,

$$Q_1 = C_1 V.$$

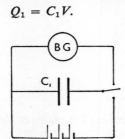

Fig. 8.27

The experiment is repeated, replacing the first capacitor by the second. Using the same nomenclature,

$$Q_2 = C_2 V.$$

$$\therefore \; \frac{C_1}{C_2} = \frac{Q_1}{Q_2} = \frac{\theta_1}{\theta_2}.$$

Energy of a parallel-plate capacitor

When the charged plates of a parallel-plate capacitor are moved towards each other the electric flux density is unchanged, since the lines of electric flux remain parallel, and hence the electric field strength is unchanged.

If one of the plates is earthed, the potential V of the other plate falls as the plates are moved together. If d is the distance between the plates, the electric intensity, V/d, remains constant.

When the plates are about to touch, the potential of the insulated plate is almost zero, since the effect of its own charge $+ Q$ is neutralized by the effect of the induced charge $- Q$ on the earthed plate.

Suppose the plates, originally at potentials of V and zero respectively, are allowed to move together until they touch,

Average potential difference $= \frac{1}{2}V$.

$\therefore$ Work done in bringing the plates together $= Q \cdot \frac{1}{2}V$

(from the definition of potential difference). This work is derived from the energy of the charged capacitor and could be utilized to lift a small weight attached to one of the plates. Since $C = Q/V$, the expression for the energy of the capacitor can be written in several forms as follows:

$$\text{Energy of charged capacitor} = \tfrac{1}{2}QV = \tfrac{1}{2}CV^2 = \frac{1}{2}\frac{Q^2}{C}.$$

EXAMPLE

Calculate the energy of a capacitor of capacitance 2 μF when charged to a p.d. of 100 V.

$$\text{Energy} = \tfrac{1}{2}CV^2$$
$$= \tfrac{1}{2} \times 2 \times 10^{-6} \times (100)^2$$
$$= 10^{-2} \text{ J.}$$

Energy of any charged conductor

By calculating the work done in charging a conductor it can be shown that the expression $\tfrac{1}{2}QV$ represents the energy of any charged conductor.

Suppose the conductor is charged by bringing up small charges from infinity. When it is at a potential V the work done in bringing up a charge δQ which is so minute that it does not appreciably change the potential, is $V\,\delta Q$. Hence to charge the conductor with charge Q

$$\text{Work done} = \int_0^Q V\,dQ$$

But $V = \dfrac{Q}{C}$,

where C = capacitance of conductor.

$$\therefore \text{ Work done} = \int_0^Q \frac{Q}{C}\,dQ$$

$$= \frac{Q^2}{2C} = \tfrac{1}{2}QV.$$

Force of attraction between the charged plates of a parallel-plate capacitor

The plates of a parallel-plate capacitor attract each other because they are oppositely charged. Suppose the force is F. This force will be constant when the plates are moved together because the electric field between the plates is constant.

When the plates, originally at a distance d apart, are allowed to move together so that they touch, the work done by the force of attraction between the plates is equal to the original energy of the capacitor, i.e.

$$Fd = \tfrac{1}{2}QV,$$

$$F = \tfrac{1}{2}Q\frac{V}{d}.$$

It might, at first sight, be thought that the force should be $Q(V/d)$, since the field between the plates is V/d. But the charge on the insulated plate may be regarded as contributing half the value of the field and the induced charge on the earthed plate as contributing the other half; the force on the charge on one of the plates is due to the field set up by the charge on the other.

EXAMPLE

Calculate the force of attraction between the plates of a parallel-plate air capacitor, of area 100 cm^2 and distant 1 mm apart, when they are charged to a p.d. of 200 V.

$$\text{Force} = \tfrac{1}{2}Q\frac{V}{d} = \frac{\tfrac{1}{2}CV^2}{d} \qquad (C = Q/V).$$

But
$$C = \frac{A\epsilon_r\epsilon_0}{d},$$

where $\epsilon_r = 1$, $\epsilon_0 = 8\cdot86 \times 10^{-12}\,\text{F m}^{-1}$.

$$\therefore\ C = \frac{100 \times 10^{-4} \times 8\cdot86 \times 10^{-12}}{1 \times 10^{-3}} = 8\cdot86 \times 10^{-11}\,\text{F}.$$

$$\therefore\ \text{Force} = \tfrac{1}{2}.8\cdot86 \times 10^{-11} \times \frac{(200)^2}{1 \times 10^{-3}}$$

$$= 1\cdot77 \times 10^{-3}\,\text{N}.$$

EXAMPLE

A parallel-plate air capacitor has a capacitance of 6 μF. Calculate the work done in separating the plates to double their original distance apart when (a) the plates are charged to a p.d. of 100 V and insulated before being separated, (b) the plates are permanently connected to a battery which maintains the p.d. at 100 V.

The capacitance of the capacitor changes from 6 μF to 3 μF when the distance between the plates is doubled. In case (a) the charge on the plates remains constant and the p.d. between them is increased. In case (b) the p.d. between the plates remains constant and the charge is decreased, some of the charge serving to charge up the battery.

(a) Charge on the plates $= CV = 6 \times 10^{-6} \times 100 = 6 \times 10^{-4}\,\text{C}$.

$$\text{Energy of capacitor} = \tfrac{1}{2}QV = \frac{1}{2}\frac{Q^2}{C}.$$

$$\text{Original energy of capacitor} = \frac{1}{2}\frac{(6 \times 10^{-4})^2}{6 \times 10^{-6}} = 3 \times 10^{-2}\,\text{J}.$$

Final energy of capacitor $= \dfrac{1}{2}\dfrac{(6\times10^{-4})^2}{3\times10^{-6}} = 6\times10^{-2}$ J.

$\therefore$ Gain in energy of capacitor $= 3\times10^{-2}$ J.

Work done in separating the plates $=$ gain in energy of capacitor
$$= 3\times10^{-2} \text{ J.}$$

(b) Original energy of capacitor $= \frac{1}{2}CV^2$
$$= \frac{1}{2}\times6\times10^{-6}\times100^2 = 3\times10^{-2} \text{ J.}$$

Final energy of capacitor $= \frac{1}{2}\times3\times10^{-6}\times100^2 = 1\cdot5\times10^{-2}$ J.

$\therefore$ Loss of energy of capacitor $= 1\cdot5\times10^{-2}$ J.

Charge flowing into battery $= (C_1-C_2)V$
$$= (6-3)10^{-6}\times100 = 3\times10^{-4} \text{ C.}$$

$\therefore$ Work done in charging battery $= 3\times10^{-4}\times100 = 3\times10^{-2}$ J.

Work done in separating the plates $=$ work done in charging battery
$$-\text{loss in energy of capacitor}$$
$$= 3\times10^{-2}-1\cdot5\times10^{-2}$$
$$= 1\cdot5\times10^{-2} \text{ J.}$$

The electrostatic voltmeter

A practical form of electrometer, commonly known as an electrostatic voltmeter, is shown in Fig. 8.28. It consists of a light metal vane which can swing inside two hollow metal quadrants. The potential difference to be measured is applied across the vane and the quadrants, when the vane is attracted into the quadrants until prevented from moving further by the controlling force provided by the weight at the bottom of the vane. The instrument must be calibrated against an absolute instrument; it can be used for measuring alternating potential differences.

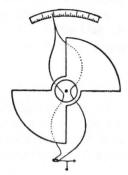

Fig. 8.28

A similar, but more sensitive electrometer, known as a quadrant electrometer, has been much used in fundamental research.

EXAMPLE

Calculate the capacitance of a sphere of radius 10 cm in air.

$$C = 4\pi a\epsilon_0$$
$$= 4\pi \times 10 \times 10^{-2} \times 8\cdot86 \times 10^{-12}$$
$$= 11\cdot1 \text{ pF.}$$

EXAMPLE

Calculate the force between two metal spheres, each of diameter 1 cm, with their centres 10 cm apart in air, when each is charged to a potential of 100 V.

We will assume that the charge on each sphere behaves as though concentrated at its centre. This is only an approximation because of the mutual repulsion of the charges on the two spheres.

$$\text{Capacitance of each sphere} = 4\pi a\epsilon_0$$
$$= 4\pi \times 0\cdot5 \times 10^{-2} \times \epsilon_0 \text{ F.}$$

$$\therefore \text{ Charge on each sphere} = CV$$
$$= 4\pi \times 0\cdot5 \times 10^{-2} \times \epsilon_0 \times 100$$
$$= 2\pi\epsilon_0 \text{ C.}$$

$$\text{Force between spheres} = \frac{Q_1 Q_2}{4\pi r^2 \epsilon_0}$$
$$= \frac{(2\pi\epsilon_0)^2}{4\pi(10 \times 10^{-2})^2 \epsilon_0}.$$

Taking $\epsilon_0 = 8\cdot86 \times 10^{-12} \text{ F m}^{-1}$,

Force between spheres $= 2\cdot8 \times 10^{-9} \text{ N.}$

E and V due to more than one charge

Potential is a scalar quantity, having either a positive or negative magnitude but no direction, and hence the potential due to two or more point charges can be summed up by straightforward algebraic addition. Electric field strength, which is equal to force per unit charge (p. 180), is a vector quantity; the resultant electric field strength due to two or more point charges is calculated by applying the parallelogram law for the addition of vectors.

Equipotential surfaces

Surfaces on which all points are at the same potential are called equipotential surfaces. Thus the equipotential surfaces in a uniform field are a set of parallel planes, equally spaced, the normals to which are lines of force. In the case of a point charge the equipotential surfaces are concentric spheres (Fig. 8.29).

It is instructive to note the analogy between electric and gravitational fields using Fig. 8.29. The circles or equipotential lines correspond to contours or lines of equal height above sea-level, while the lines of

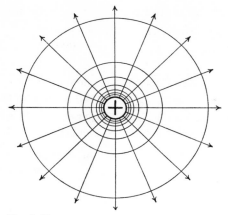

Fig. 8.29

electric force correspond to lines of greatest gravitational slope. We can ascend a hillside by walking at right angles to the contours up the line of the greatest slope or we can zigzag taking more gentle slopes. Similarly, the potential gradient at an angle to a line of electric force is less than along the line of force. Again, the force along the vertical in a gravitational field is $W = mg$; along a line of force in an electric field the force is $F = QE$.

The surface of a conductor must be an equipotential surface; if there were a difference of potential between two points of a conductor, charge would flow until the difference disappeared. Hence there can be no component of an electric field parallel to the surface of a conductor, and all lines of electric force must everywhere be perpendicular to the surface. No line of electric force can begin and end on the same conductor, since there must be a fall of potential down a line of electric force.

Summary

The electric potential at a point, in volts, is the work done, in joules, in bringing up a charge of 1 coulomb from infinity to the point.

Capacitance is measured in farads.

$$C = \frac{Q}{V}.$$

Charge resides on the surface of a conductor and tends to accumulate on the more curved parts.

The total induced charge is always equal to the inducing charge.

Relative permittivity of a material, ϵ_r

$$= \frac{\text{capacitance of a capacitor with the material as dielectric}}{\text{capacitance of a similar capacitor with a vacuum as dielectric.}}$$

The electrophorus and the Van de Graaff generator are commonly used in the laboratory for producing electrostatic charges.

Electric field strength measured in volts per metre

$$= \text{potential gradient,}$$

i.e. $E = -\dfrac{dV}{dx}.$

Also   Electric field strength (newtons per coulombs)

$$= \text{force per unit charge}$$

i.e. $E = \dfrac{F}{Q}.$

Electric flux through an area A where the Electric Intensity is E

$$= EA = \frac{Q_1 + Q_2 + \cdots}{\epsilon} \quad \text{when } A \text{ surrounds } Q_1, Q_2, \text{ etc.}$$

Relative permittivity, $\epsilon_r = \dfrac{\epsilon}{\epsilon_0}.$

$$\epsilon_0 = \frac{1}{\mu_0 c^2} \simeq \frac{10^{-9}}{36\pi} \text{ F m}^{-1}.$$

Permittivity is measured in farads per metre.

Capacitance of parallel plate capacitor $= \dfrac{\epsilon A}{d}.$

Energy of a charged capacitor $= \tfrac{1}{2}QV.$

Attraction between plates of capacitor $= \tfrac{1}{2}Q\,\dfrac{V}{d}.$

Capacitors in parallel $C = C_1 + C_2 + C_3,$ etc.

Capacitors in series: $\dfrac{1}{C} = \dfrac{1}{C_1} + \dfrac{1}{C_2} + \dfrac{1}{C_3},$ etc.

Capacitance of sphere $= 4\pi\epsilon r$.

Force between point charges $= \dfrac{Q_1 Q_2}{4\pi\epsilon d^2}$.

Questions

Take $\epsilon_0 = \dfrac{10^{-9}}{36\pi}$ farad per metre unless otherwise instructed.

1. Describe a gold-leaf electroscope. How would you show that the charge on an ebonite rod, when rubbed with fur, is equal and opposite to that on the fur?

2. Define *potential* at a point and *capacitance* of a conductor (or capacitor).
A capacitor requires a charge of 2 μC to raise its potential by 100 V. What is its capacitance?

3. Explain, in terms of electrostatic induction, why a charged body attracts an uncharged body.
A charged polythene rod is brought up to an uncharged pith-ball suspended by a silk thread. The pith-ball is first attracted to the rod, touches it and is then repelled. Explain.

4. One end of an uncharged, insulated, cigar-shaped conductor A is brought near to a positively charged conducting sphere B. Show by a diagram the charges on A.
A wire, connected at one end to the cap of an electroscope, is held by an insulating rod and its other end is moved along A from one end to the other. Describe and explain what happens.
Draw diagrams showing the potential of B and its neighbourhood (*a*) before, (*b*) after A is brought near.

5. Describe and explain how you would charge a gold-leaf electroscope negatively by induction. Draw diagrams showing the charges at three stages of the process.
Explain why the presence of a positive charge in a conductor should not be tested by means of a negatively charged electroscope.

6. Describe the experiment known as Faraday's ice-pail, and show what important conclusions can be drawn from it. Do you consider it an *exact* proof of these conclusions?
Being supplied with an insulated metal sphere charged with 0·01 μC, a small and a large calorimeter each with an insulating stand and handle, how would you produce (*a*) a charge of $-0·02$ μC, (*b*) a charge of $+0·02$ μC.   (B adapted)

7. A gold-leaf electroscope is placed on a block of polythene and the cap, or knob, is connected by a wire with the metal case of the instrument. A small electric charge is given to the cap. The connecting wire is then removed by means of insulating tongs. Finally the cap is touched with the finger. Describe and explain the sequence of events.   (D)

8. How would you investigate experimentally the charge distribution over the surface of a conductor? Describe and explain the action of points on a charged conductor and give two practical applications of the effect.

9. Two insulated metal plates, A and B, are placed parallel to each other and about 1 inch apart. Each is connected to the cap of an electroscope. State and explain the indications of the electroscopes (a) when A is given a positive charge, (b) when B is then earthed, (c) when a slab of polythene is then introduced between A and B, while B remains earthed. Explain how the capacitance of A is affected in each case.

10. Describe an electrophorus and, with the aid of diagrams, explain how it is used to give positive charges to an insulated conductor. Why is the plate of the electrophorus at a high potential when lifted from the base although it has previously been earthed?

Why is it possible to obtain a succession of charges from an electrophorus? What is the source of the energy associated with these charges?   (N)

11. How would you show (a) that there are two kinds of electric charge, 'positive' and 'negative', and (b) that a charge from a voltaic cell can be made to give a deflection on a gold-leaf electroscope?

Describe how, given an electrometer and accessory apparatus, you would determine the sign and compare the magnitude of the charges on two small insulated bodies. You can assume that the charge on a glass rod rubbed with silk or a cellulose acetate strip rubbed with wool is positive.   (O & C)

12. Define *electric flux* or *displacement* and state how it is related to electric field strength in free space.

A parallel plate capacitor consists of two square plates of side 30 cm at a distance of 2 mm apart in air. One of the plates is earthed and the other plate is given a charge of $1 \cdot 20$ μC which raises its potential to 3000 V. Calculate the permittivity of air.

13. Deduce an expression for the capacitance of a parallel-plate capacitor.

It is required to construct a capacitor of capacitance $0 \cdot 20$ μF consisting of plane metal sheets separated by mica ($\epsilon_r = 6$) of thickness $0 \cdot 0010$ cm. What must be the area of each of the metal sheets?

If the capacitor is connected to the poles of a 4 V battery, calculate its charge.

14. A capacitor of capacitance $0 \cdot 400$ μF is charged by an accumulator of e.m.f. $2 \cdot 12$ V and discharged through a ballistic galvanometer. If the throw in the galvanometer is $20 \cdot 0$ divisions, find the coulomb sensitivity of the galvanometer.

15. Find the total capacitance of three capacitors of 1, 2 and 4 μF when they are connected (a) in parallel, (b) in series.

If the applied p.d. is 100 V, find the charge on each capacitor, and the p.d. across it, in both (a) and (b).

16. Two capacitors A and B are charged by means of a certain battery when they are connected (a) in series, (b) in parallel. On discharging them through a ballistic galvanometer the kicks are respectively 4 and 18 divisions. Determine the ratio of their capacities.

17. Give an account of the action of an electrostatic capacitor, and show that the energy of such a capacitor is $\frac{1}{2}CV^2$, where C is its capacity and V the potential difference between the electrodes.

To what potential must a capacitor of capacity 40 μF be charged to have the same energy as a fully charged 2 V, 10 A hr accumulator?   (O & C)

18. A capacitor of capacitance 2 V has one of its plates earthed and the other plate is charged to a potential of 120 V. The charged plate is connected to one plate of another capacitor, of capacitance 1 μF, the other plate being earthed, and when the connection is made a small spark passes.

Calculate the energy of the capacitors (a) before, (b) after being connected, and account for the energy of the spark.

19. Define *electric field strength*. Obtain an expression for the force exerted on a charge Q in a field of electric intensity E.

The force exerted by an electric field on a charge of 2 μC is 4×10^{-4} N. What is the electric field strength of the field?

20. Explain whether it is possible

(a) for the potential at a point to be zero while the electric field strength at the point is not zero;

(b) for the electric field strength at a point to be zero while the potential is not zero;

(c) for charge to be distributed in such a way that both the potential and the electric field at a point are zero.

21. Two plane electrodes, 0·6 cm apart, within an electron tube (or valve), are maintained at a difference of potential of 100 V. Determine the electric field strength in the vacuum between the electrodes.

An electron is a point charge of magnitude $1·60 \times 10^{-19}$ C and its mass is $9·1 \times 10^{-31}$ kg. How long would it take an electron, starting at rest from one electrode, to reach the other? (L adapted)

22. Two brass plates are arranged horizontally, one 2 cm above the other, and the lower plate is earthed. The plates are charged to a difference of potential of 6000 V. A drop of oil with an electrical charge of $1·60 \times 10^{-19}$ C is in equilibrium between the plates so that it neither rises nor falls. What is the mass of the drop? (O & C adapted)

23. Two small, gilded, spherical pith-balls, each of mass 0·004 g, are suspended from the same point by threads each of length 10 cm. Assuming the balls to be equally charged and to remain in equilibrium with the balls 2 cm apart, calculate the charge on each.

24. (a) Calculate the electric field strength in air at distances of 1, 2, 5 and 10 cm from a point charge of 0·005 μC. Plot a graph of electric field strength and distance from the charge.

(b) Do the same for potential.

25. A parallel-plate air capacitor has its plates 2 cm apart. If a slab of glass 2 cm thick is placed between the plates it is found necessary to increase their distance apart by 1·6 cm in order to restore the capacity to its original value.

Explain this, giving the underlying cause of the effect produced by the dielectric. Calculate the dielectric constant of the glass. What proportion of the total energy of the electric field will be in the glass in the final arrangement of the capacitor? (N)

26. A capacitor consisting of two parallel plates each of area 100 cm², distant 1 cm apart with a slab of glass ($\epsilon_r = 10$), 0·8 cm thick, filling most of the space

between them. One of the plates is earthed and the other is given a charge of $0.030 \, \mu C$. Calculate the potential of the insulated plate and also its electrical energy.

The slab of glass is removed. What are now the potential and electrical energy of the insulated plate. Where did the extra energy come from?

27. Mica can store electrical energy up to about $1500 \, \mathrm{J} \, \mathrm{m}^{-3}$. If its relative permittivity is 6.0, how thick should be the sheets of mica in a parallel plate capacitor for p.d.'s up to 500 V?

28. A thundercloud and the earth can be regarded as a plane parallel capacitor. Taking the area of the thundercloud as $50 \, \mathrm{Mm}^2$, its height above the earth as 1 km, and its potential as 100 kV, calculate the energy stored in the capacitor.

29. Find the capacitance of a capacitor consisting of a sphere and a concentric spherical conductor, of radii 5 and 10 cm respectively, separated by air.

30. What is the capacitance in farads per kilometre of a cable consisting of a cylindrical copper core of diameter 3.0 mm, which is surrounded by rubber insulation ($\epsilon_r = 3.0$) of thickness 5.0 mm and encased on the outside by lead sheathing?

31. Calculate the capacitance of the earth, taking its radius as $6.4 \, \mathrm{Mm}$.

32. Two spheres of radii 6 and 2 cm respectively, at a considerable distance apart, are connected by a wire. How will a charge of $1 \, \mu C$ distribute itself, and what will be the surface densities, in $\mu C \, \mathrm{m}^{-2}$, on the two spheres?

33. Calculate the work done in charging a sphere of radius 5 cm to a potential of 1000 V.

34. Assuming that the charge on an isolated conducting sphere behaves as though concentrated at its centre, find an expression for the electric field strength at a point outside the sphere.

Given that air becomes ionized (i.e. conducting) at an electric field strength of $3 \times 10^6 \, \mathrm{V} \, \mathrm{m}^{-1}$, find the greatest charge that can be placed on a sphere of radius 1 cm in air.

35. A capacitor consists of two parallel plates of 10 cm radius separated by an air gap of 1 mm. Calculate the work done in separating the plates to a distance of 1 cm when (*a*) the plates are connected to a 100 V battery, (*b*) the plates are charged to a p.d. of 100 V and the plates are insulated before being separated.

(C S)

36. What is meant by the terms (*a*) *potential* and (*b*) *field strength* in electrostatics? State whether each quantity is a scalar or a vector.

Write down the law which gives the force between two point charges Q_1 and Q_2 at a distance r apart and use it to derive the electric field strength and the potential due to a point charge Q at a distance x from it.

The points A, B and C form an equilateral triangle of side z. Point charges of equal magnitude Q are placed at A and B. Find the electric field strength and the potential at C due to these charges when (i) both charges are positive and (ii) the charge at A is positive and the charge at B is negative.   (O & C 1968)

37. Explain what is meant by the terms *electric field strength* and *electric potential*.

Four infinite conducting plates A, B, C and D, of negligible thickness, are arranged parallel to one another so that the distance between adjacent plates is 2 cm. The outer plates A and D are earthed and the inner ones B and C are maintained at potentials of $+20$ V and $+60$ V respectively. Draw a graph showing how the potential between the plates varies as a function of position along a line perpendicular to the plates. What is the magnitude and direction of the electric field between each pair of adjacent plates?

By using Gauss's theorem, or otherwise, determine the charge density on plates B and C (in coulombs per square metre). If the inner plates were isolated and then connected together, what would be their final potential?   (O & C 1970)

38. Explain why the electric field strength close to the surface of a charged conductor is always at right angles to the surface of the conductor.

Describe the Van de Graaff generator and explain its mode of operation.

A model Van de Graaff generator has a hollow sphere of radius 0.5 m mounted on a column of resistance 6×10^9 Ω. The belt conveys charge to the sphere at the rate of 10^{-6} A.

(a) What is the steady potential of the sphere?

(b) What is the surface density of charge on the sphere?

(c) What is the electric potential gradient close to its surface?

(Take the value of ϵ_0 to be 8.85×10^{-12} F m^{-1}.)   (O 1971)

39. State Gauss' Theorem in electrostatics. Prove the theorem for the particular case of a point charge at the centre of a spherical surface.

Assuming the general validity of the theorem, obtain the expression for the electric field strength at a point very close to a conductor on which the surface density of charge is σ.

What is the mechanical force experienced by a small element δS of the surface of the conductor?

An insulated metal plate of area 0.2 m^2 is held at a distance of 3×10^{-3} m from a parallel earthed conducting plate; it carried a charge 10^{-6} C. What is the mechanical force on the insulated plate? In which direction does it act? What is the p.d. between the surfaces?

Find the mechanical force on the insulated plate and the potential difference between the plates when the separation is increased from 3×10^{-3} m to 3×10^{-2} m.

(Take the value of the permittivity of free space, ϵ_0, to be 8.85×10^{-12} F m^{-1}.)
                                                                             (O 1968)

40. Define *capacitance*. Obtain from first principles an expression for the capacitance of a parallel-plate capacitor.

Describe the experiments you would perform in order to determine how the capacitance of such a capacitor varies with the area of overlap of the plates, and the separation of the plates.

A parallel-plate air capacitor has plates each of area 10^{-2} m^2 separated by 2×10^{-3} m. If the p.d. between the plates is 200 V, calculate (a) the charge, (b) the stored energy, and (c) the force between the capacitor plates.

(Take the value of ϵ_0 to be 8.85×10^{-12} F m^{-1}.)   (O 1969)

9   Alternating current

The current supplied to most of the buildings in this country comes from the Grid and it is alternating, with a frequency of 50 Hz. Alternating current is used in preference to direct current because it is easier to generate, and much easier to transform from one voltage to another. Transmission of electric power is much more economical at a high voltage than at a low one (p. 14, 90); hence alternating voltage is transformed up before entering the Grid from the generating station, and down again before entering houses or factories.

Alternating current is generated when a coil is rotated uniformly in a magnetic field:

$$E = E_0 \sin 2\pi ft \quad \text{(p. 96)},$$

where E represents the e.m.f. at time t, E_0 the peak value of E, and f the frequency of the a.c. Using a similar nomenclature for current,

$$I = I_0 \sin 2\pi ft.$$

Root-mean-square current

To measure an a.c. an instrument must be selected whose indication is independent of the direction of the current, for example, a moving-iron or hot-wire instrument, or a moving-coil instrument fitted with a rectifier. The mean value of an a.c. is zero; what then does the instrument measure?

The main purpose of a.c. is to transfer energy from one point to another. The *effective* or *virtual* value of an a.c. is taken as *the value of d.c. which produces the same amount of heat in the same time.*

Since the heat produced is proportional to the square of the current (heat $= I^2 Rt$), the effective value of an a.c. is the square root of the mean value of I^2; this can be shown to be equal to $I_0/\sqrt{2}$ for a sinusoidal wave as follows, where I_0 is the peak value:

$$I^2 = I_0^2 \sin^2 2\pi ft$$
$$= \tfrac{1}{2}I_0^2 (1 - \cos 4\pi ft).$$

But the average value of cos $4\pi ft$, for any whole number of cycles, is zero.

$\therefore$ Mean value of $I^2 = \frac{1}{2}I_0^2$.

$\therefore$ Root mean square value of $I = I_0/\sqrt{2}$.

Thus a sinusoidal current with a peak value of 10 A has a r.m.s. value of 7·07 A.

Similarly, the virtual, effective or r.m.s. value of E is $E_0/\sqrt{2}$. Alternating current supplied from the Grid at 240 V has an effective p.d. of 240 V; the peak value is $240 \times \sqrt{2} = 340$ V.

Flow of a.c. through a capacitor

If a capacitor of fairly high capacitance, say 2 μF, is connected through a 15-watt lamp to the 240 V-a.c. mains (Fig. 9.1), the lamp lights. The a.c. is flowing, apparently, through the capacitor which consists of two sets of plates separated by an insulating dielectric. What is happening is that the capacitor is being charged, discharged, charged in the opposite

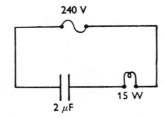

240 V

2 μF

15 W

Fig. 9.1

direction, and discharged, 50 times per second (the frequency of the a.c.), and the charging and discharging currents flow through the lamp.

If the capacitor is replaced by one of considerably smaller capacitance, for example, 0·2 μF, the lamp does not light because the charging and discharging currents are not sufficiently high. It is clear that the resistance, to which the capacitance is equivalent, is less the larger the capacitance. It is also less the larger the frequency of the a.c.

Capacitor connected to d.c. supply

Further information about the flow of a.c. in a circuit containing a capacitor can be obtained by considering what happens when a capacitor is connected to a source of d.c. Suppose that a 2 μF capacitor is connected through a milliammeter to a 100 V battery. When the switch in

Fig. 9.2 is closed the milliammeter shows a momentary deflection while the capacitor is being charged.

The charging process can be slowed down, and rendered visible, by

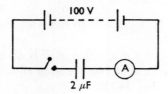

Fig. 9.2

including in the circuit a high resistance, for example 2 MΩ, and replacing the milliammeter by a microammeter (Fig. 9.3). The charging current is then smaller and lasts for a much longer time, about 10 s. It is found that the current is initially high, while the capacitor is uncharged,

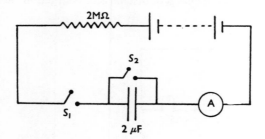

Fig. 9.3

and becomes smaller as the capacitor becomes charged. The microammeter does not respond immediately to the current because of the inertia of its moving parts. The charging process can be better observed by first short circuiting the capacitor by closing the switch S_2 in Fig. 9.3 when the maximum current flows through the microammeter. On opening S_2, the current gradually decreases (Fig. 9.4) as the capacitor becomes charged and as its increasing p.d. opposes that of the battery.

Lead of the current in a capacitive circuit

Just as the maximum current flows immediately a capacitor is connected to a source of d.c., while the capacitor is uncharged, so the maximum current flows immediately a capacitor is connected to a source of a.c. The result is that the current leads the applied e.m.f., and the lead is

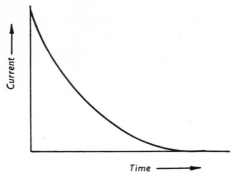

Fig. 9.4

$\frac{1}{2}\pi$ or $\frac{1}{4}$ cycle if the circuit has negligible resistance (Fig. 9.5, ignoring the shaded areas).

This current lead can be demonstrated by a beautifully simple experiment, making use of electrolysis. A piece of blotting paper is soaked in potassium sulphate and phenolphthalein and is laid on a brass plate.

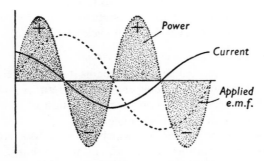

Fig. 9.5 Circuit possessing capacitance only.

If the brass plate is connected to the positive pole of a 12 V battery and a piece of thick copper wire with a smooth rounded end is connected to the negative pole of the battery and drawn across the blotting paper, a purple line is formed on the blotting paper, caused by the electrolysis of the potassium sulphate; the release of caustic potash at the cathode (i.e. the moving wire) causes the phenolphthalein to change from colourless to purple. If a source of a.c. is used, by connecting the brass plate and moving wire to the 240 V-mains through a 15-W lamp (to prevent a short-circuit), a series of purple dashes, with colourless spaces between, are formed on the blotting paper; the moving wire is alternately the cathode and the anode, 50 times per second.

Two circuits are now connected simultaneously to the 240 V mains (Fig. 9.6 (*a*)), one containing resistance only (a 15-W lamp), and one containing capacitance only (a 2-μF capacitor). Two wires, side by side,

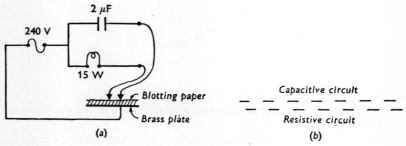

Fig. 9.6

are drawn across the blotting paper and it is found that the purple dashes formed by the capacitive circuit lead those formed by the resistive circuit by approximately $\frac{1}{4}$ of the length occupied by one purple dash and one colourless interval, i.e. by $\frac{1}{4}$ cycle (Fig. 7.6 (*b*)).

Power absorbed in a capacitive circuit

When a capacitor is connected across a.c. mains it alternatively absorbs power from the mains while it is being charged and returns power to the mains while it discharges. There is no net absorption of power in the circuit if the resistance of the circuit is negligible.

The power absorbed in the circuit, which is the product of the instantaneous values of the applied e.m.f., E, and the current, I, is represented in Fig. 9.5. Since E and I are both zero once in each half-cycle, the power curve has a frequency double that of E or I. The shaded areas represent the product of power and time and hence the energy absorbed in the circuit. In the lower shaded loops, marked $-$, the current is flowing in a direction opposite to the applied e.m.f., and hence is giving up energy to the source of supply. It can be seen from the figure that there is a transference of equal amounts of energy to and fro between the source of supply and the capacitor in each half-cycle.

Effect of inductance in an a.c, circuit

When a current in a coil is changing, the magnetic flux links with the turns of the coil and gives rise to an induced back e.m.f. The phenomenon is called self-induction (p. 103) and the coil is said to possess inductance.

The great importance of inductance in the case of alternating current may be demonstrated by the apparatus in Fig. 9.7. A coil is connected to a source of 12 V a.c. through a 12 V, 24 W lamp. The coil will probably

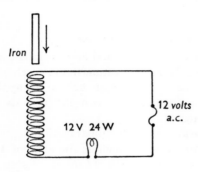

Fig. 9.7

have sufficient inductance, even with an air core, to prevent the lamp from reaching its full brightness. But when an iron rod is inserted into the coil the lamp dims owing to the increased inductance, and consequent increased back e.m.f. in the coil. The coil and iron rod are known as a choke.

Effect of inductance on the growth of a d.c.

The current in a coil, connected to a d.c. supply, takes time to grow and decay because the e.m.f. due to self-induction always opposes the change of the current. In a highly inductive circuit, such as the field windings of a dynamo or motor, the current may take many seconds to grow.

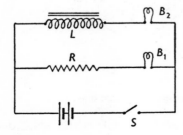

Fig. 9.8

The slow growth of the current in an inductive circuit can be demonstrated with the apparatus in Fig. 9.8. Two 2·5 V bulbs, B_1 and B_2, are connected to a 4 V battery through a resistance and a high inductance respectively. When the switch S is depressed B_1 lights first and B_2 afterwards. The inductance L can conveniently take the form of the two coils of an electromagnet, the poles of the magnet being connected by an iron yoke.

Lag of the current in an inductive circuit

The current in an inductive circuit grows and decays slowly, inductance having similar effect on current as mass or inertia on acceleration. This results in the current in an a.c. circuit lagging behind the applied e.m.f. and the lag is $\frac{1}{2}\pi$ or $\frac{1}{4}$ cycle if the circuit has negligible resistance (Fig. 9.9, ignoring the shaded areas).

The lag can be demonstrated with the apparatus in Fig. 9.6 (a), replacing the capacitor by an inductor, such as the coils of an electromagnet fitted with a yoke. Referring to Fig. 9.6 (b), the dashes caused by an inductive circuit would lag behind the dashes caused by the resistive circuit by approximately $\frac{1}{4}$ cycle.

Power absorbed in an inductive circuit

In a circuit consisting of a coil of negligible resistance connected across a.c. mains, the current, when it is increasing, is always in the same direction as the applied e.m.f. and the circuit absorbs energy from the mains (Fig. 9.9). When the current is decreasing its direction is always opposite to that of the applied e.m.f. and energy is returned to the mains.

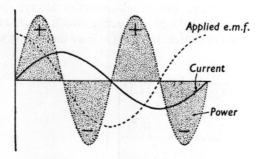

Fig. 9.9 Circuit possessing inductance only.

The only opposition to the applied e.m.f. is the e.m.f. due to self-induction and the current continuously adjusts itself so that these two e.m.f.s are always equal and opposite. The e.m.f. due to self-induction is in the opposite direction to the current when the current is increasing and in the same direction as the current when the current is decreasing. It can be seen from Fig. 9.9 that the net power, represented by the shaded areas marked + and −, is zero. Hence a purely inductive circuit absorbs no power.

Inductance and capacitance in a circuit

Capacitance causes the current to lead and inductance causes it to lag, and hence they can be made to annul each other's effects. In Fig. 9.10 a 75-W lamp is connected in series with the coils of an electromagnet

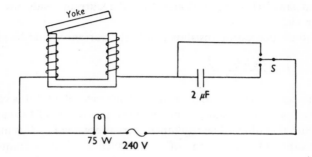

Fig. 9.10

(providing inductance) and with a 2 μF capacitor, to the 240 V a.c. mains. S is a two-way switch which enables the capacitor to be short-circuited. When the capacitor is short-circuited, lowering the yoke across the poles of the electromagnet, which increases the inductance, has the effect of a choke and dims the lamp. With the capacitor in the circuit, lowering the yoke causes the lamp to brighten; this is because increasing the inductance annuls some of the capacitance and hence increases the current.

Capacitance in an a.c. circuit

We will now represent mathematically the facts which we have discussed concerning a circuit consisting of a capacitor connected by wires of negligible resistance to a source of alternating e.m.f. Let the capacitance of the capacitor be C farads and the alternating e.m.f. be represented by $E_0 \sin 2\pi ft$ volts. Let the charges on the plates at any instant be Q and $-Q$ coulombs.

$$\textbf{P.d. between plates} = \frac{Q}{C}.$$

But p.d. between plates = applied e.m.f.

$$\therefore \ \frac{Q}{C} = E_0 \sin 2\pi ft,$$

$$I = \frac{dQ}{dt} = 2\pi fCE_0 \cos 2\pi ft$$

$$= 2\pi fCE_0 \sin (2\pi ft + \tfrac{1}{2}\pi).$$

The term $\sin (2\pi ft + \tfrac{1}{2}\pi)$ indicates that the current leads the applied e.m.f. by $\tfrac{1}{2}\pi$.

In a circuit possessing resistance only the current would be given by

$$I = \frac{E_0 \sin 2\pi ft}{R}.$$

Hence the term $1/2\pi fC$ is analogous to resistance, R, and it is called the *capacitive reactance*, usually abbreviated to reactance, of the capacitor; it is measured in ohms. The reactance decreases when f or C is increased. The reactance of a capacitor of capacitance $2\,\mu\text{F}$, at a frequency of 50 Hz, is

$$\frac{1}{2\pi . 50 . 2 \times 10^{-6}} = 1590 \ \Omega.$$

At a frequency of 10^6 Hz, a typical radio frequency, the reactance is $0 \cdot 08 \ \Omega$.

Inductance in an a.c. circuit

Consider a circuit consisting of a coil of negligible resistance connected to a source of alternating e.m.f. Let the inductance of the coil be L henrys and let the a.c. flowing through it be represented by $I = I_0 \sin 2\pi ft$. Since there is no resistance in the circuit the only opposition to the applied e.m.f. is the e.m.f. due to self-induction, $L(dI/dt)$ (p. 103).

$$\therefore \ \textbf{Applied e.m.f.} = L\frac{dI}{dt}$$

$$= LI_0 2\pi f \cos 2\pi ft$$

$$= 2\pi fLI_0 \sin (2\pi ft + \tfrac{1}{2}\pi).$$

The term $\sin(2\pi ft + \frac{1}{2}\pi)$ indicates that the applied e.m.f. leads the current by $\frac{1}{2}\pi$.

In a circuit possessing resistance only the applied e.m.f. would be $RI_0 \sin 2\pi ft$. Thus the term $2\pi fL$ is analogous to R. It is called the *inductive reactance*, usually abbreviated to reactance, and it is measured in ohms. The reactance increases with the frequency, f. The reactance of a coil of inductance 1 henry, for a frequency of 50 Hz, is $2\pi.50.1 = 314\ \Omega$. For a frequency of 10^6 Hz the reactance is $6.28\ M\Omega$.

Resistance and inductance in an a.c. circuit

Consider a circuit containing both resistance and inductance (Fig. 9.11). At first sight it might be thought that the total 'resistance' of the circuit is $R + 2\pi fL$; this, however, does not allow for the phase lag of the current due to the inductance.

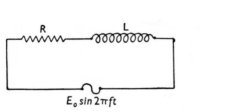

$E_0 \sin 2\pi ft$

Fig. 9.11

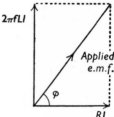

Fig. 9.12

The problem is solved most simply by means of phasors. The applied e.m.f. must supply, at any instant, RI volts to drive the current through the resistance, in phase with the current; it must also supply $2\pi fLI$ V to overcome the back e.m.f. due to self-induction, with a phase lead of 90° over the current. The applied e.m.f. is the result of these two phasors (Fig. 9.12); it must therefore lead the current by an angle ϕ, known as the phase difference or *angle of lag*, where

$$\tan\phi = \frac{2\pi fIL}{RI} = \frac{2\pi fL}{R}.$$

From Fig. 9.12

$$\text{Applied e.m.f.} = \sqrt{\{R^2 + (2\pi fL)^2\}}I.$$

The term $\sqrt{\{R^2 + (2\pi fL)^2\}}$, analogous to resistance, is called the *impedance* of the circuit and is measured in ohms.

Power factor

The component of the applied e.m.f., $2\pi fLI$, is often called the wattless component because no power is dissipated in the inductance. Power is, however, dissipated, in the form of heat, in the resistance. Fig. 9.13 represents the power curve; the current lags behind the applied e.m.f. by less than 90°, and hence the positive loops of the power curve are larger than the negative loops.

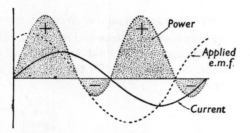

Fig. 9.13. Circuit possessing resistance and inductance.

The power dissipated in the resistance is I^2R. From Fig. 9.12,

$$IR = E \cos \phi, \qquad \text{where } E = \text{applied e.m.f.}$$

$$\therefore \text{ Power dissipated} = IE \cos \phi.$$

The term $\cos \phi$ is called the *power factor*.

The true power dissipated in the circuit is thus $I_v E_v \cos \phi$ (where I_v and E_v are virtual or root mean square values): it is measured in watts. The apparent power is $I_v E_v$ and is measured in VA. The power factor is the ratio of the true power to the apparent power.

The wattmeter

An instrument for measuring the true power in an a.c. circuit, known as a wattmeter, is shown in Fig. 9.14 (*a*). It consists of a coil, *B*, pivoted between two fixed coils *A*, *A*, and controlled by hair springs. The connections of the three coils to the load, in which the power is to be measured, are shown in Fig. 9.14 (*b*). The same current passes through the coils *A*, *A*, as through the load (since the current through *B* and the high resistance is very small); the same p.d. is applied across *B* and the high resistance as across the load. Thus the force on the moving coil is proportional to the product of the current through, and the p.d. across

MICHAEL FARADAY   LORD KELVIN

MICHAEL FARADAY (1791–1867), *one of the greatest of experimental physicists, became first the laboratory assistant and then the successor to Sir Humphrey Davy at the Royal Institution. His electrical researches began with the experiment, attempted earlier without success by Wollaston, in which a wire carrying a current rotated round a magnetic pole – the first electric motor. Ten years later, after repeated failures to convert magnetism into electricity, he discovered electromagnetic induction and made the first dynamo. His concepts of lines and tubes of force, and of the action of the dielectric, provided a new and revolutionary viewpoint which led Maxwell to formulate the electromagnetic theory of light. He discovered the laws of electrolysis and made numerous advances in electrostatics. He was a deeply religious man, wholly absorbed in his scientific investigations and indifferent to money. He refused a knighthood and the Presidency of the Royal Society.*

WILLIAM THOMSON, LORD KELVIN (1824–1907), *became professor of natural philosophy at Glasgow university at the early age of twenty-two, and held this chair for fifty-three years. For most of his life he was a busy man of affairs, as well as a professor, and much of his original scientific work was done in the railway train between Glasgow and London. Lord Rutherford said of him that he 'combined to an extraordinary degree the quality of great theoretical insight with the power to realize his ideas in a practical form'. Two of his finest theoretical achievements were the theory of electrical oscillations in a circuit containing resistance, inductance and capacitance, and the foundation (with Carnot and Clausius) of thermodynamics. His inventions included the quadrant electrometer, the direct-reading ammeter, improved forms of ship's compass and sounding apparatus, and the siphon recorder for use with the Atlantic cable.*

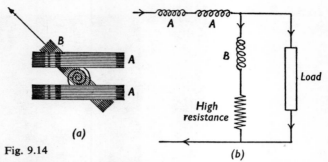

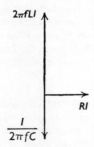

(a)

Fig. 9.14

(b)

the load, i.e. the power in the load. By detaching the high resistance in series with B and replacing it by a different resistance the range of the instrument can be changed.

The apparent power can be obtained by using a separate ammeter and voltmeter. From the values of the true and apparent power the power factor can be calculated.

Circuit containing resistance, inductance and capacitance

Figure 9.15 represents the components of the applied e.m.f. for a circuit containing resistance, inductance and capacitance. The impedance of the circuit must therefore be

$$\sqrt{\left\{R^2+\left(2\pi fL-\frac{1}{2\pi fC}\right)^2\right\}}.$$

$2\pi fLI$

RI

$\dfrac{I}{2\pi fC}$

Fig. 9.15

Electrical resonance

Since inductance causes lag and capacitance causes lead it is possible to adjust their values in a circuit so that they neutralize each other. Only the resistance is then effective in limiting the current which is a maximum. Since inductance is proportional to frequency and capacitance is inversely proportional to frequency, the two can be equated only for a particular frequency. This is known as the *resonant frequency* and at this frequency the circuit is said to be in a state of resonance. For resonance,

$$2\pi fL = \frac{1}{2\pi fC},$$

$$\therefore f = \frac{1}{2\pi\sqrt{(LC)}}.$$

If L is in henrys and C in farads, then f is in hertz. The equation is of great importance in radio.

Demonstration of resonance

The frequency of the oscillator in Fig. 9.16 is set to about 200 Hz and the lamps L_s and L_1 are seen to light, indicating that at this low frequency the current prefers to pass through the inductive arm of the circuit. The frequency is then gradually increased until a frequency is reached (about 280 Hz) when L_s goes off, but L_1 and L_2 are on. This

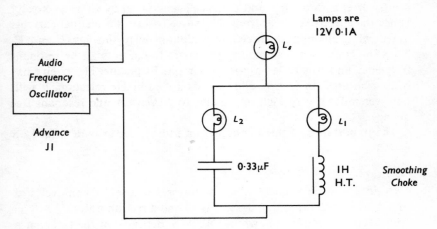

Fig. 9.16

marks the resonance point of the parallel circuit with the circulating current much larger than the supply current. If the frequency is raised still higher L_s comes on again but L_1 goes out, showing that at high frequencies the current prefers to pass through the capacitive arm.

Acceptor and rejector circuits

Figure 9.17 (*a*) and (*b*) represent the application of an alternating e.m.f. to an inductance and a capacitance connected in series and in parallel respectively. The resonant frequency is approximately the same for both circuits but they behave differently.

In circuit (*a*) the current is a maximum and the impedance is a minimum at the resonant frequency; the circuit is called an *acceptor* circuit.

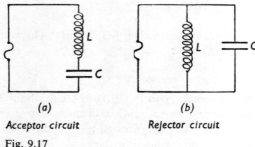

(*a*)   (*b*)

Acceptor circuit   Rejector circuit

Fig. 9.17

Suppose the resistance of circuit (*b*) is zero. The source of supply sends currents through L and C which lag and lead by 90° respectively; hence the effective current from the source is zero, since equal currents are flowing in opposite directions simultaneously through L and C. Thus the circuit acts as a complete barrier to currents at the resonant frequency and it is called a *rejector* circuit. In practice the circuit has some resistance, of course, and there is not complete rejection. It will, however, offer a very high impedance to currents at the resonant frequency.

The importance of these circuits in a radio receiver is described on p. 265.

The metal rectifier

A rectifier is an instrument which suppresses one set of half-cycles of an a.c. so that the resulting current is in one direction only.

One type of rectifier, shown in Fig. 9.18, depends on the fact that a current flowing from copper oxide to copper experiences a low resistance, but when flowing in the opposite direction it experiences a very high resistance. The rectifier consists of a number of specially treated copper disks, bolted together with lead washers, metal spacers and large metal fins to dissipate heat. One side of each of the copper disks has been subjected to special heat treatment to form a thin layer of copper oxide. This type of rectifier was often used for a high-tension supply to mains-operated radio receivers.

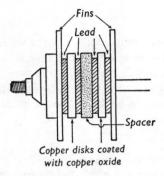

Fig. 9.18

An alternative but quite popular method of rectification was to use thermionic vacuum diode valves (see p. 231). These days, however, the reader is far more likely to find semiconductor diodes in use (see p. 278).

Bridge-rectifier method of adapting moving-coil instruments to measure a.c.

Figure 9.19 represents a method of connecting four rectifiers to a d.c. moving-coil instrument to measure a.c. The arrangement allows both half-cycles of the a.c. to pass in the same direction through the instrument which, however, measures the rectified mean and not the r.m.s. value of the current.

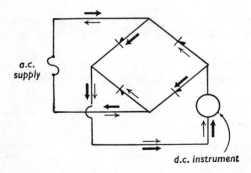

Fig. 9.19

Summary

An alternating e.m.f., E, can be represented by

$$E = E_0 \sin 2\pi ft.$$

Similarly, $I = I_0 \sin 2\pi ft.$

The effective or virtual value of a sinusoidal a.c. is the value of a d.c. which produces the same amount of heat in the same time and it is equal to the root mean square current, $I_0/\sqrt{2}$, where I_0 is the peak value.

Capacitance in an a.c. circuit causes the current to *lead* the applied e.m.f.; *inductance* causes the current to *lag* behind the applied e.m.f.

Questions

1. Figure 9.20 represents a method of finding the frequency of the a.c. mains. The a.c. is passed through a sonometer wire, PQ, which runs between the poles of a strong horseshoe magnet. The length of the wire is adjusted until it vibrates to the maximum extent. Explain fully why the wire vibrates.

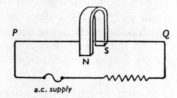

Fig. 9.20

Suppose the length of the wire resounding to the a.c. is 77·4 cm, while the length which resounds to a tuning fork of frequency 128 Hz is 30·3 cm (no current then passing through the wire but the tension remaining unaltered). Calculate the frequency of the a.c.

2. What is meant by the root mean square value of an alternating current? A moving-iron ammeter in an a.c. circuit reads 2·0 A. What is the maximum or peak value of the current?

3. A voltmeter placed across a.c. mains reads 200 V. What is the maximum p.d. across the mains? If the frequency of the a.c. is 50 Hz, what expression represents the mains voltage?

4. Compare and contrast the effect in the case of d.c. and a.c. when a resistance, an inductance and a capacitor are placed in the circuit.   (O & C)

5. What is meant by an a.c.? How is such a current obtained? A coil containing inductance and resistance is supplied with an alternating e.m.f. of constant amplitude. Draw a rough graph showing how the power delivered alters as the frequency of alternation increases.   (OS)

6. Explain descriptively, in terms of the flow of electric charge, what happens when a capacitor is connected across the a.c. mains, and justify the statement that in this case the *current leads the e.m.f.*

Explain further why the amplitude of the current increases with the capacity of the capacitor; and, for a given capacitor, with the frequency of the alternating e.m.f.

Draw any circuit that makes use of the property of a capacitor to pass a.c. and to block d.c., and discuss its uses.   (O & C)

7. An 8 μF capacitor is placed across the 200 V r.m.s. a.c. 50 Hz electrical supply. Calculate the r.m.s. current which flows in the circuit. What is the peak value of the voltage across the capacitor?   (CS)

8. A capacitor is connected to a source of alternating voltage. Explain why the current flow in the leads is a maximum when the value of the applied voltage is zero.

What is the effect of introducing a resistor in series with the capacitor? In a circuit of this type it is found that when the r.m.s. voltage of the alternating source is 250 V, the r.m.s. voltage across the capacitor and resistor are 150 and 200 V respectively; explain this observation.   (O & C)

9. A moving-iron ammeter in series with an air-cored solenoid reads 7 A when connected across the 210 V d.c. mains and 4·2 A when connected to the 210 V 50 Hz a.c. mains. Calculate the self-inductance of the solenoid, and the power consumed by the coil when on the a.c. circuit. The resistance and inductance of the ammeter may be neglected.   (O & C)

10. Write a short account of resonance, illustrating your answer by examples taken from light and electricity.   (CS)

11. Describe a cathode-ray tube.

An alternating e.m.f. is applied to a resistance AB, a capacity BC and an inductance CD connected in series. The X-plates are connected across AB. What is the pattern on the screen when the Y-plates are connected (a) across AB also, (b) across BC, (c) across CD, (d) across AD and the capacity changed from a very small to a very high value?   (OS)

12. Find the impedance of a telephone receiver of resistance 100 Ω and inductance 0·002 H (a) at 1000 Hz, (b) at 100 000 Hz.

13. Find the magnitude and phase of the current which flows when an alternating p.d. of r.m.s. value 200 V and frequency 50 Hz is applied to the ends of the following:

(a) a coil of inductance 0·2 H and negligible resistance;
(b) a capacitor of capacitance 80 μF;
(c) a coil of inductance 0·2 H and of resistance 20 Ω;
(d) a capacitor of capacitance 80 μF in series with a resistance of 20 Ω;
(e) a coil of inductance 0·2 H and resistance 20 Ω in series with a capacitor of capacitance 80 μF.

14. A coil of inductance 0·2 mH and negligible resistance is in series with a capacitor of capacitance 0·000 5 μF. Find the resonant frequency of the circuit.

15. A p.d. with a frequency of 50 Hz is applied to a coil of inductance 4 H and resistance 200 Ω. What is the power factor of the coil?

16. An a.c. of 2 A and frequency 100 Hz flows through a coil of inductance 2 mH and resistance 4 Ω. What is the p.d. across the coil, the power absorbed in the coil, and the power factor?

17. A coil, of resistance 20 Ω and inductance 0·5 H, is connected in series with a capacitor, of capacitance 40 μF, to a 230 V, 50 Hz supply. Determine (a) the current, (b) the voltage drop across the coil and capacitor.

18. A circuit consists of a resistance, an inductance and a capacitance in series. The resistance of the circuit is 200 Ω, and the inductive and capacitive reactances are each 200 Ω at a frequency of 50 Hz. Taking different values of the frequency from 0 to 100 Hz, plot rough graphs of (1) resistance and frequency, (2) inductive reactance and frequency, (3) capacitive reactance and frequency, (4) impedance and frequency, (5) phase difference and frequency.

19. Show how a moving-coil meter can be used to measure alternating currents with the help of a suitable rectifier bridge.

When such a rectifier bridge incorporating a moving-coil ammeter is connected in a direct-current circuit in series with a moving-iron meter, the readings of the two instruments agree. In 50 Hz alternating-current circuits, the moving-coil meter consistently reads about 10 per cent lower than the moving-iron meter. Explain carefully the reason for this. Which of the two readings do you regard as having the greater significance? (O 1969, part)

20. While a straight wire of length L is moving across a uniform magnetic field of flux-density B, the e.m.f. induced between the ends of the wire is BLv units, where v is the component of the velocity of the wire perpendicular both to its length and to B. Starting from this fact, show how a single-phase sinusoidal alternating voltage can be generated by a rectangular coil rotating steadily in a uniform magnetic field. What factors determine (i) the peak value, (ii) the frequency, of the voltage generated?

Explain what is meant by the root-mean-square (r.m.s.) value of an alternating current or voltage, and show that for a sinusoidal variation of the quantity the r.m.s. value is $1/\sqrt{2}$ times the peak value.

Figure 9.21 shows the variation with time of (a) a sinusoidal alternating voltage, (b) an alternating voltage after half-wave rectification, (c) an alternating voltage after full-wave rectification. State the r.m.s. value of the voltage in each case.

(O 1968)

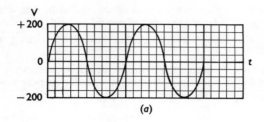

(a)

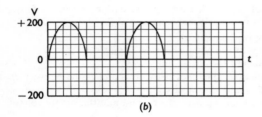

(b)

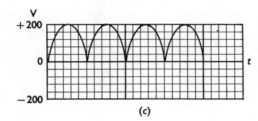

(c)

Fig. 9.21

21. The unit of magnetic flux, the weber or volt-second, is defined using the equation $E = -d\Phi/dt$, where E stands for e.m.f. and Φ stands for magnetic flux. Outline the experimental basis for this equation.

A simple alternator when rotating at 50 revolutions per second gives a 50 Hz alternating voltage of r.m.s. value 24 V. A 4 Ω resistance R and a 0·01 H inductance L are connected in series across its terminals.

(i) Assuming that the internal impedance of the generator can be neglected, find the r.m.s. current flowing, the power converted into heating, and the r.m.s. potential difference across each component.

(ii) Draw a vector diagram showing the relative phases of the applied voltage and the potential differences across R and L.   (O 1970)

22. Define *capacitance*. Describe a simple direct method of measuring a capacitance which is known to be of the order of one microfarad.

Explain how an alternating current can flow in a series circuit containing a capacitor.

A sinusoidal voltage of frequency f and peak value V is applied to a series circuit having resistance R and capacitance C. Obtain formulae for the peak value and the phase lead of the current in the circuit.   (O 1968)

23. Explain how a sinusoidal alternating voltage may be generated by a rotating coil, and obtain an expression for the voltage developed at time t.

For the circuit of Fig. 9.22, carrying alternating current of frequency 50 Hz, calculate, correct to two significant figures,

   (*a*) the reactance of the inductor L,
   (*b*) the impedance of the circuit between A and C,
   (*c*) the r.m.s. value of the circulating current,
   (*d*) the r.m.s. potential difference between A and B and between B and C,
   (*e*) the power dissipated in the circuit,
   (*f*) the phase difference between the current and the supply voltage.

                                                                (O 1969)

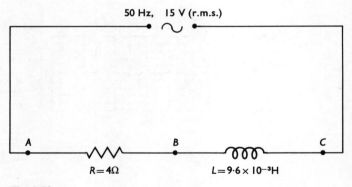

Fig. 9.22

10 Thermoelectric, photoelectric and thermionic effects

Seebeck and Peltier effects

In 1821 Seebeck discovered that if the junctions of two different metals in a circuit are kept at different temperatures, a small current flows. The phenomenon is known as the thermoelectric or Seebeck effect. If the junctions of the iron–copper thermocouple in Fig. 10.1 (*a*) are maintained at 273 and 373 K respectively, the thermo-e.m.f. is of the order

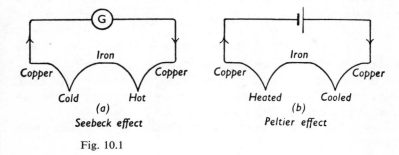

(*a*)
Seebeck effect

(*b*)
Peltier effect

Fig. 10.1

of 1 mV; the conventional current flows from copper to iron through the hot junction.

In 1834 Peltier discovered the inverse effect. If a current passes through a circuit containing two different metals, heat is generated at one junction and absorbed at the other (Fig. 10.1 (*b*)). The effect occurs whether the current is driven by an external cell or whether it is generated by the thermocouple itself. Thus, when the Seebeck effect occurs, heat is continually absorbed at the hot junction and developed at the cold junction; likewise when the Peltier effect occurs it tends to set up a Seebeck e.m.f. which opposes the current.

Experimental demonstration

The Seebeck effect is easily demonstrated. It is necessary only to heat the junction of pieces of copper and iron wire with a match, the other ends of the wires being connected to the terminals of a fairly sensitive galvanometer. The cold junction in this case is at the galvanometer.

The Peltier effect can be demonstrated with a bismuth tellurium junction such as that available from the W. B. Nicolson Division of Baird & Tatlock. Connect the terminals marked 'galvanometer' to a lamp-and-scale galvanometer, or 'amplified' galvanometer having a sensitivity of a few cm per μA. Touch the thermocouple for an instant and note that the slight heating sends the spot off the scale. Now connect the bismuth tellurium junction to a single 2 V lead–acid accumulator or 'Nife' cell, with the polarity as marked. About 12 A are taken, so short and fairly thick leads are needed. For a short time the cooling effect dominates and the spot on the galvanometer is deflected in the opposite direction, but the Joule heating effect soon takes over, and the spot then moves in the same direction as when the thermocouple was touched with a finger. Now reverse the connections to the Bi–Te junction and note that the galvanometer indicates heating only. Do not leave on for more than a few seconds.

Laws of intermediate metals and of intermediate temperatures

The following two laws have been established experimentally:

(1) If A, B and C are three different metals, the thermoelectric e.m.f. of the couple AC is equal to the sum of the e.m.f.'s of the couples AB and BC over the same temperature range.

It follows that the junctions of a thermocouple may be soldered without affecting the e.m.f.

(2) The e.m.f. of a thermocouple, with junctions at temperature θ_1 and θ_3, is the sum of the e.m.f.'s of two couples of the same metals with junctions at θ_1 and θ_2, and at θ_2 and θ_3, respectively.

It follows that when a galvanometer is connected to a thermocouple, as in Fig. 10.1 (a), the e.m.f. is independent of the temperature of the galvanometer.

Measurement of thermoelectric e.m.f.

The e.m.f. of a thermocouple may be measured by means of a potentiometer. Since the thermo-e.m.f. is of the order of a millivolt it is necessary to arrange that the potential drop across the potentiometer wire is only slightly more than this by putting a resistance R in series (see Fig. 10.2). Then if E is the e.m.f. of the cell, and r is the resistance of the potentiometer wire, the potential drop across potentiometer wire $= \{r/(R+r)\}E$. The internal resistance of the cell has been ignored, but R must be several hundred ohms and hence no appreciable error is introduced.

The junctions of the thermocouple are maintained at the desired temperatures, for example, those of melting ice and boiling water, while the balance point K is found.

Then e.m.f. of thermocouple $= \dfrac{AK}{AB} \dfrac{rE}{R+r}$.

The value of E may be found with moderate accuracy by means of a voltmeter. Alternatively, the potentiometer can be calibrated with a standard cell as described on p. 29.

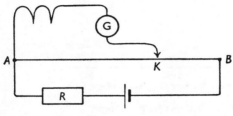

Fig. 10.2

Variation of thermo-e.m.f.'s with the temperatures of the junctions

The variation of the e.m.f. of a thermocouple when the temperature of one junction is kept constant and the temperature of the other junction is varied, may be investigated by the method just described.

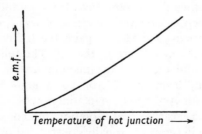

Fig. 10.3. Iron–constantan thermocouple.

Figures 10.3 and 10.4 show graphs of e.m.f. and hot-junction temperature, the cold junction being maintained at 273 K, for iron–constantan and iron–copper thermocouples. The former is part of a parabola but is appreciably straight for the range 273–373 K, and is therefore suitable for measuring temperature. The latter is an almost perfect parabola; the temperatures of the hot junction at which the e.m.f. is a maximum and afterwards zero are called, respectively, the neutral temperature and the temperature of inversion.

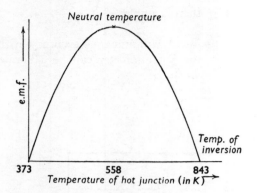

Fig. 10.4. Iron–copper thermocouple.

Thermoelectric series

For small values of temperature difference between the junctions, the metals can be arranged in a thermoelectric series as follows: antimony, iron, zinc, copper, silver, lead, aluminium, mercury, platinum–rhodium, platinum, nickel, constantan (60% copper, 40% nickel), bismuth.

When two of these metals are combined to form a thermocouple, the conventional current flows from the one earlier in the list to the other across the cold junction; thus the current flows from *a*ntimony to *bis*-muth through the *c*old junction (note the italicizing of *a*, *b*, *c*, which suggests a method of memorizing the fact). Again, the thermo-e.m.f. is greater the farther apart the metals are in the list. The following are several thermo-e.m.f.'s in millivolts when the junctions are maintained at 273 and 373 K: antimony–bismuth 12, iron–constantan 5, copper–constantan 4, platinum with 10% rhodium–platinum 0·64.

Thermopiles designed to measure heat radiation and consisting of a number of thermocouples in series are usually constructed of antimony and bismuth; iron and constantan also give a comparatively high e.m.f. and they are sometimes used. For measuring high temperatures the alloy platinum–rhodium and platinum are employed because of their high melting-points.

The Thomson effect

A thermocouple may be regarded as a reversible heat engine, absorbing heat at one junction, rejecting it at the other, and generating electrical energy equivalent to the difference in these two quantities of heat. William Thomson (later Lord Kelvin) applied the theory of heat engines to a thermocouple and deduced that the law of variation between the e.m.f. and the temperatures of the junction should be the same whatever the nature of the metals used. This is contrary to experiment. He therefore predicted another thermal effect, known as the *Thomson effect*, which can be stated as follows: *if a temperature gradient exists in a conductor an electric potential gradient also exists*. As an analogy, the pressure of the air in a glass tube heated at one end and cooled at the other is the same throughout, but the density of the air is greater at the cool end than at the hot end. Similarly, the pressure of the electron gas in an unevenly heated conductor is the same throughout, but its density varies; hence there is a p.d. in the conductor.

The Seebeck effect can be explained fully in terms of the Peltier and Thomson effects. By Peltier effect is here meant the reversible effect at the junction. An absorption or evolution of heat equivalent to π joules, when 1 coulomb of electricity flows, is equivalent to an e.m.f. of π volts (π is called the Peltier coefficient). At the two junctions of a thermocouple there are Peltier e.m.f.'s in opposite directions, the e.m.f. at the hot junction being greater than that at the cold. In the two metals there are temperature gradients and hence Thomson e.m.f.'s which are small compared with the Peltier e.m.f.'s. The algebraic sum of the Peltier and Thomson e.m.f.'s is equal to the Seebeck e.m.f.

It should be noted that a thermoelectric e.m.f., which is of the order of a millivolt, is a phenomenon quite distinct from a contact potential, which is of the order of a volt. The former is a transport effect and the latter an equilibrium effect. The contact p.d.'s at the two junctions of a thermocouple act in opposite directions.

Photoelectric effect

When light falls on certain metals electrons are emitted from them; this is called the photoelectric effect.

The effect was discovered accidentally by Hertz in 1887 while performing his experiments on electromagnetic waves. He found that a spark jumped more readily across the gap between the metal knobs of his receivers when the knobs were illuminated by ultra-violet radiation. Hallwachs studied the phenomenon; he found that when light from an arc falls on a negatively charged zinc plate connected to a gold-leaf electroscope, the plate and electroscope become discharged. If the plate and electroscope are positively charged there is no leakage of the charge. Subsequently it was shown by J. J. Thomson and Lenard, independently, that the action of the light caused the metal to emit electrons; a positive charge on the metal prevents the electrons from escaping.

It is now known that most metals emit electrons when illuminated by ultra-violet light of suitable frequency; the alkali metals, lithium, sodium, potassium, rubidium and caesium emit electrons under the action of visible light as well as of ultra-violet light.

Einstein's photoelectric equation

An experimental study of the photoelectric effect revealed a surprising fact. The velocity of the emitted electrons is rigorously independent of the intensity of the light. If the intensity of the light is increased the number of electrons emitted is increased but their velocity is unchanged. The velocity of the electrons does depend, however, on the frequency of the light – the greater the frequency, the greater the velocity. Thus ultra-violet light is more effective than visible light in causing electrons to be emitted, because it has a greater frequency.

In 1901 Planck had put forward a revolutionary theory to account for the observed radiation of heat and light from hot bodies. He assumed that energy is radiated in the form of discrete units called quanta, and that each quantum has energy hf, where h is a universal constant now called the Planck constant, and f is the frequency of the radiations. The quantum theory was revolutionary because it challenged, and seemed completely to contradict, the well-established wave theory of radiation.

In 1905 Einstein proposed an explanation of the photoelectric effect on the quantum theory. He suggested that when a quantum of energy hf (now also called a photon) strikes a metal, part of its energy is used in liberating an electron, and the rest in giving the electron kinetic energy; the photon is annihilated. The work required to liberate an electron is ϕe, where ϕ is the electronic exit work function or work function and e is the charge of the electron.

Suppose m is the mass of the electron and v its velocity of emission; then

$$hf = \phi e + \tfrac{1}{2}mv^2.$$

This is one of the most famous and significant equations of the physics of this century, and it gained for Einstein a Nobel prize. Its significance lies in the fact that it is one of the main foundations on which the quantum theory rests.

Millikan's experimental verification of Einstein's equation

Figure 10.5 represents diagrammatically the apparatus used by Millikan in a classic series of experiments by which he verified Einstein's equation and made determinations of h and ϕ.

The plate A is made of the metal whose photoelectric emission is to be investigated. It is illuminated by light of measured frequency f, from a source L. The plate B is given a negative potential V by means of the potentiometer P, which is just sufficient to prevent the electrons emitted by A from reaching it. When no electrons reach B from A the electrometer E will register zero deflection.

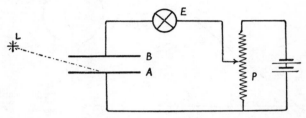

Fig. 10.5

The work done against the stopping potential is equal to the kinetic energy of the electrons, i.e.

$$Ve = \tfrac{1}{2}mv^2.$$

Thus $\quad Ve = hf - \phi e \quad$ (from Einstein's equation).

If f_0 = frequency of the radiation which just enables the electron to escape with zero velocity, called the threshold frequency,

$$hf_0 = \phi e,$$
$$\therefore \ Ve = hf - hf_0.$$

Thus, if Einstein's theory is correct, V should be a linear function of f. Millikan found that this was the case (Fig. 10.6). The value of h can be determined from the slope of the graph, knowing e, since $h = Ve/(f-f_0)$; Millikan obtained $h = 6 \cdot 56 \times 10^{-34}$ J s. The value of ϕ can be found from the intercept on the axis of f since $\phi = hf_0/e$.

A series of values of ϕ for different metals is given in the table below.

Electronic exit work functions (volts)

Caesium	1·87–1·96	Zinc	3·57
Rubidium	2·16–2·19	Copper	4·16
Potassium	2·24	Tungsten	4·54
Lithium	2·28	Silver	4·74
Sodium	2·46	Platinum	6·30

Since a contamination of the surface by a thin film of oxide has a considerable effect on the emission of electrons, Millikan enclosed the plates B and A in a vacuum and, just before illuminating A, removed a thin shaving from its surface with a knife which could be operated from outside the vacuum. B was made of copper and an allowance had to be made for the considerable contact potential difference between A and B; the stopping potential was taken as the difference between the potentiometer reading and the contact potential difference.

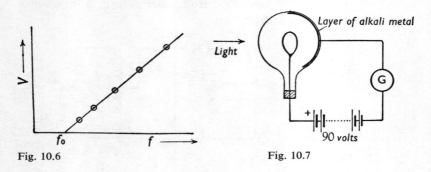

Fig. 10.6   Fig. 10.7

The photoemissive cell

The photoemissive cell consists of a glass bulb provided with two electrodes. The cathode consists of a coating of an alkali metal such as sodium, potassium or caesium on part of the inside surface of the glass; the anode is a loop of platinum wire (Fig. 10.7). When light falls on the coating of alkali metal, electrons are emitted and these can be attracted to the anode if the latter is raised to a suitable positive potential. Hence a small current flows through the circuit in Fig. 10.7. The bulb may be evacuated or contain an inert gas such as argon. It is essential that oxygen should be excluded since once the photoelectric film has become oxidized it loses much of its emissive power.

The vacuum photoemissive cell has, because of its size and high voltage requirement, been largely replaced by semiconductor devices. It may still be found, however, in all but recent sound film projectors.

Sound films

The sound track of a film is made as follows. The sound vibrations are converted by a microphone (similar in principle to a telephone transmitter) into corresponding fluctuating electric currents. The currents are

Fig. 10.8. (*a*) Variable density sound track (from *Gone with the Wind*).
(*b*) Variable area sound track (from *2001: A Space Odyssey*).

amplified and made to open the jaws of a narrow slit through which light shines on to a strip at the edge of the film. Loud sounds open the slit wider than soft sounds, and hence the sound track has a variable density (Fig. 10.8 (*a*)).

In another method the varying currents cause the length of an oscilloscope trace to vary, hence giving a variable area sound track as opposed to the variable density track (Fig. 10.8 (*b*)).

When the film is projected, light from an 'exciter' lamp is passed through the sound track into a photoelectric cell (see Fig. 10.9), thus giving rise to varying currents which are converted back to sound waves by a loud speaker. For the picture projection it is necessary to jerk the

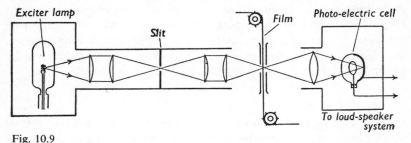

Fig. 10.9

film forward at the rate of 24 frames per second, but the sound apparatus requires perfectly steady motion past the photoelectric cell. To enable these conflicting requirements to be satisfied there must be a free length of film between the two parts of the machine and, in consequence, the sound track is printed 25 frames ahead of the pictures to which it corresponds.

Thermionic effect

When a metal is heated it emits electrons; this is called the thermionic effect.

An electron cannot escape from the surface of a metal unless its kinetic energy, $\frac{1}{2}mv^2$, exceeds a certain value: viz. $\frac{1}{2}mv^2 \geqslant \phi e$, where ϕ is the electronic work function, and e is the charge of the electron. As the temperature of the metal is raised, the kinetic energies of the electrons increase and hence more electrons can escape.

The effect is similar to the increase in the evaporation of a liquid with temperature. The heat of evaporation of the electrons, which is ϕe electron-volts per electron, is analogous to the latent heat of vaporization of a liquid. This suggests a method for the determination of ϕ which was employed successfully by Davisson and Germer in 1922.

The electrons emitted by a hot metal accumulate in in the space round the metal until there is a dynamic equilibrium between the electrons outside and inside the metal, similar to the equilibrium of a vapour over a liquid. If, however, the electrons are drawn away from the metal by a plate at a positive potential, electrons will continue to escape from the metal. Davisson and Germer measured the electrical energy which it was necessary to supply to a tungsten filament to maintain its temperature at a fixed value, (1) when a plate near to the filament was kept at a positive potential to draw off the electrons, and (2) when the plate was kept at a negative potential to prevent the emission of electrons. The difference in the energy inputs to the filament in the two cases gave the heat of evaporation of the electrons; the number of electrons escaping in the former case could be calculated from the current flowing between the filament and the positive plate. The values of ϕ obtained were in good agreement with those obtained from the photoelectric effect.

The diode

In 1883 Edison inserted another electrode into an ordinary carbon filament lamp, an arrangement similar to Fig. 10.10, and discovered that

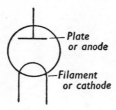

Fig. 10.10

a current will flow through the vacuum inside the bulb if the extra electrode is made positive with respect to the filament, but not if it is negative. The phenomenon is called the Edison effect, and it is now realized that the hot filament of the lamp is emitting electrons; the extra electrode attracts the electrons if it has a positive potential, but repels them if it has a negative potential.

In 1904 Dr (afterwards Sir Ambrose) Fleming used the device for radio reception and hence invented the first wireless valve. It is called a valve because it allows a current to pass through it in one direction only; alternatively, it is called a diode because it has two electrodes.

The way in which the current through a diode varies with the p.d. between the anode and the filament can be investigated by the circuit in Fig. 10.11. The filament is heated by the 'low-tension' supply and the positive potential of the anode is varied by means of a 'high-tension' variable supply. The current, measured with a milliammeter, flows through the diode, through the milliammeter, through the high-tension supply, and through the circuit of the low-tension supply.

The graph of the current against anode potential is called the characteristic of the diode and is shown in Fig. 10.12. The graph flattens out at the top and the maximum current is called the *saturation current*; all the electrons emitted by the filament are then proceeding to the anode. It is

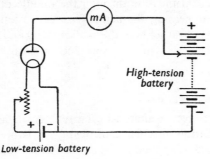

Low-tension battery

Fig. 10.11

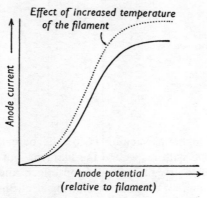

Fig. 10.12

natural to ask why all the electrons are not attracted to the anode as soon as a small positive potential is applied to the anode. The reason is that the space round the cathode is filled with a cloud of electrons called the *space charge*; these set up an electric field in opposition to that of the anode. The middle part of the graph is comparatively straight and over this portion the change in current is proportional to the change in anode potential. The effect of raising the temperature of the filament, by slightly increasing the heating current, is shown in the dotted graph in Fig. 10.12.

The filament of a valve is made of tungsten, which is usually coated with oxides of thorium or of barium and strontium, thereby increasing its electron emission by as much as twenty times; an oxide-coated filament is run at a much lower temperature than a non-coated filament and continues to emit for a long period. Usually, in a 'valve' mains radio, valves with indirectly heated cathodes are used. The cathode takes the form of an oxide-coated metal tube through the middle of which, insulated from it, runs a heating filament. This construction gives a cathode system which has a rather high thermal capacity and so prevents any fluctuation in the emission of electrons due to a variation in the heating current.

The diode as a rectifier

The diode may be used to rectify alternating current. A single diode produces half-wave rectification in which half of the a.c. is not used (Fig. 10.13). An alternating p.d. is applied to the anode of the diode by means of a transformer. The method of heating the filament is not shown in the

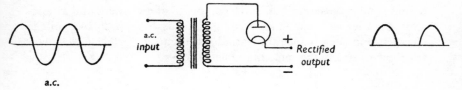

Fig. 10.13. Half-wave rectification.

figure; an extra small secondary coil on the transformer is used. A positive, conventional current can flow through the diode in one direction only, from anode to filament (actually electrons flow in the opposite direction). Thus the output current must flow from the terminal connected to the filament, marked + in the figure.

Figure 10.14 represents full-wave rectification in which both halves of each cycle of the a.c. are utilized. The valve is a double-diode containing one filament and two anodes. The secondary of the transformer has a

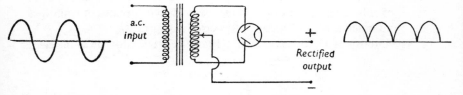

Fig. 10.14. Full-wave rectification.

centre-tap which is connected to the negative output terminal. As in half-wave rectification the filament must be connected to the positive output terminal since a positive conventional current can flow through the valve only from one of the anodes to the filament, and not in the reverse direction.

The secondary of the transformer may be regarded as two separate coils, each of which is used only for alternate half-cycles. The two circuits operate separately like Fig. 10.13.

The output current of a rectifier is pulsating; it can be smoothed by

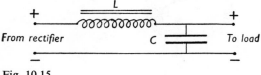

Fig. 10.15

a filter similar to that in Fig. 10.15 composed of a choke L and a capacitor C. The back e.m.f. induced in L tends to oppose the voltage changes, and thus reduce the variations in current, while C acts as a reservoir to steady the remaining fluctuations.

Fig. 10.16

The triode

In 1907 Lee de Forest introduced a third electrode into the valve and so produced the triode (Fig. 10.16), which became as important and fundamental in wireless transmission and reception as the telescope in astronomy or the microscope in biology. There have been about 1000 types of valve in use, but the triode is the prototype because it embodies the fundamental idea of a control electrode or grid, G.

The interior of a triode is shown in Fig. 10.17. The cathode or filament

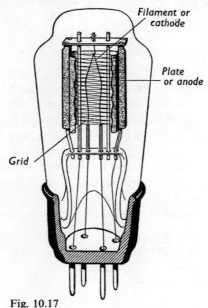

Fig. 10.17

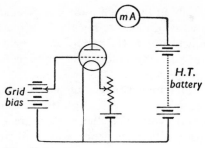

Fig. 10.18

is a coated tungsten wire and round it is the grid in the form of an open spiral of nickel wire; the anode or plate surrounds both. The grid is much more effective than the anode in controlling the current through the valve because it is nearer to the filament; a change of voltage on the grid causes a bigger change in the electrostatic field near to the filament than an equal change in the voltage of the anode. Because of their high velocity the electrons do not follow the electric field lines but shoot straight through the grid spaces which cover a much greater area than the wires.

If the grid voltage is kept constant and the anode voltage, V_A, is varied, the graph of V_A and of current between filament and anode, I_A, is similar to that for a diode (Fig. 10.12).

The most interesting and useful graphs are those for V_G, the voltage of the grid, and I_A, when V_A is kept constant. These are called the *mutual characteristics* of the valve.

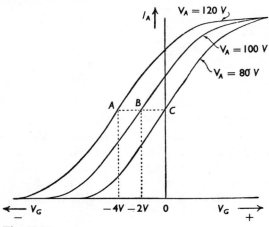

Fig. 10.19

Figure 10.18 shows the circuit by which these characteristics may be obtained. A constant potential difference, say 100 V, is maintained between the anode and filament by the HT supply and the current in the milliammeter is read as the voltage on the grid is varied by means of the grid-bias supply.

Three characteristics, for anode voltages of 80, 100 and 120, are shown in Fig. 10.19. Consider points A, B, C on the three characteristics, and suppose that the grid voltages at these points are -4, -2 and 0 V respectively. A change can be made from A to B, or from B to C, without altering the current, by a simultaneous increase in the grid voltage of 2 and a decrease in the anode voltage of 20. Thus a change of 2 V on the grid has the same effect on the current as a change of 20 V on the anode. The valve is said to have an *amplification factor* of $\frac{20}{2} = 10$.

Alternatively we may define the amplification factor, μ, as

$$\frac{\text{change in anode voltage to produce a small change in anode current}}{\text{change in grid voltage to produce the same change in anode current}},$$

i.e. $\qquad \mu = \dfrac{\delta V_A}{\delta V_G}.$

The use of the triode in receivers and oscillators will be discussed in Chapter 11.

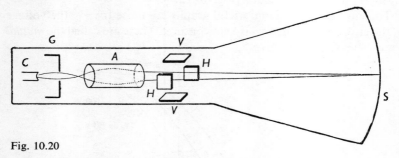

Fig. 10.20

The cathode-ray oscilloscope

The cathode-ray oscilloscope is an instrument whose fundamental principle is the deflection of an electron beam by an electric (or magnetic) field (see Fig. 10.20) and it may be used for showing the waveform of a.c. and of sound waves, for radiolocation, for television and for numerous other purposes. It has become an extremely valuable and important instrument.

A stream of electrons from a heated filament, C (the cathode), is focused on to a fluorescent screen, S, in an evacuated glass bulb, and may be deflected horizontally and vertically by p.d.'s between the plates HH and VV respectively. Since the inertia of the electrons is very small they respond instantly to rapid changes in voltage between the plates HH and VV. The perforated metal cylinder, G, is maintained at a negative potential with respect to C and hence repels the electrons. By varying its potential the flow of electrons, and hence the brightness of the spot on the screen, can be controlled; as it is a control electrode it is called a grid or Wehnelt cylinder. The hollow cylindrical anode A is maintained at a positive potential of 500 V or more with respect to C and serves to accelerate and focus the stream. A is connected to S so that the electrons travel at uniform speed after leaving A. The screen S is coated with a thin opalescent layer of a salt such as calcium tungstate, zinc phosphate, or zinc orthosilicate, which fluoresces a blue, green or bluish white colour when struck by fast-moving electrons.

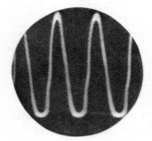

Fig. 10.21. A trace obtained with a cathode-ray oscilloscope showing the wave form of a.c. mains.

The time-base

To show how any alternating voltage varies with time, as for example in Fig. 10.21, the spot of light on the screen of the cathode-ray oscilloscope must be made to move at a uniform speed from left to right, and to return very rapidly to its starting point on the left, for another crossing. At the same time, the varying voltage to be 'graphed' is made to move the spot vertically.

To produce the horizontal motion, a voltage which varies in a 'sawtooth' pattern must be applied (Fig. 10.22). There exist many refined circuits for this purpose, but a *very* simple one is shown in Fig. 10.23. If the resistance R is high, charge will flow onto C exponentially, but the early part of the exponential will be near enough linear. When the p.d. across C reaches the 'striking' voltage of the neon N, the capacitor C

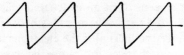

Fig. 10.22

discharges in a very short time to the extinction voltage of the neon, which is near enough zero. The X plates HH being connected effectively across C, the spot moves as required.

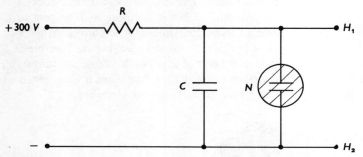

Fig. 10.23

One use of the oscilloscope is shown in Fig. 10.24.

Television

Television transmission consists in transforming the variations in brightness of a picture into corresponding fluctuating electric currents, amplifying the currents and broadcasting them as electromagnetic waves; reception consists of amplifying the fluctuating currents from the receiving aerial and converting them back into a picture.

Both in the television camera and in the receiver there is an electron beam which 'scans' the picture. One time base scans the beam from left to right, just as in the previous section, while another scans it vertically. In the new British 625 line standard the beam is scanned $312\frac{1}{2}$ times from left to right for each vertical scan (or frame) and a total of 50 vertical scans are made per second. Actually alternate sets of horizontal lines are scanned in successive frames and a complete picture contains 625 lines. A complete picture is covered 25 times a second but the alternate line scanning, with one full size half-detail picture every $\frac{1}{50}$ of a second, helps to prevent flicker.

The technique of decomposing a picture by scanning is the essence of television. In a camera each point on the picture, scanned in this way, is made to yield a voltage depending on its brightness. In a receiver the

Fig. 10.24. B/B_0 curves of mild steel (left) and stalloy (right) obtained with a cathode-ray oscilloscope (compare with Fig. 5.8). The method employed is as follows. The iron under test is in the form of a closed ring; round it are wrapped a primary coil which carries the magnetizing a.c. (proportional to B_0) and a secondary coil in which the induced e.m.f. is proportional to the rate of change of the flux, i.e. to dB/dt. The primary and secondary coils are connected through a suitable circuit to the two pairs of plates of the cathode-ray oscilloscope so that the varying p.d.s are proportional to B_0 and B.

beam strength is varied by this voltage so that each point is given an appropriate brightness and the picture reconstructed.

Television camera

Figure 10.25 shows the principle of the vidicon camera tube. Three external coils are needed, namely, the alignment coil, the focussing coil, and the horizontal (and vertical) deflection coils (electrostatic deflection needs high voltages and magnetic deflection coils are used instead.) It is

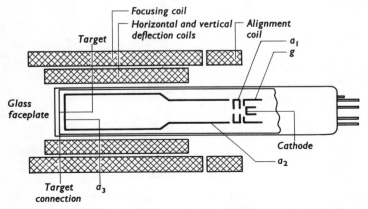

Fig. 10.25. Schematic electrode and coil arrangement of a vidicon camera tube.

helpful to consider the tube as consisting of three sections: the electron gun, the scanning section, and the target section.

The electron gun contains a thermionic cathode, a grid g to control the beam current, and an anode a_1 which accelerates the electrons and releases them in a fine beam.

In the scanning section, the beam from a_1 enters the space enclosed by the long cylindrical electrode a_2. The electrons are brought to a point focus on the target by the combined action of a fixed axial magnetic field produced by the focusing coil, and the field between a_2 and a_1, which is adjusted by altering the voltage on a_2.

At the far end of a_2 is a metal mesh a_3, connected to a_2. It produces a uniform decelerating field in front of the target, which is at a relatively low potential. This focused beam is made to scan the target by two pairs of deflection coils.

The alignment coil close to the hole in a_1 is used to ensure that the electrons from a_1 are properly aligned with the axial magnetic field of the focusing coil.

Figure 10.26 shows the target part of the vidicon, with a simple lens arrangement. The target consists of an optically flat faceplate, which has a transparent conductive film on its inner surface. This film is connected to an external target-electrode ring. On the conductive film is deposited a thin layer of photoconductive material; it has a high specific resistance

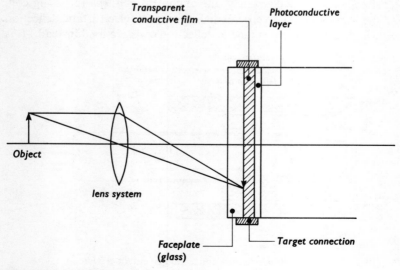

Fig. 10.26. Vidicon target.

in dark, and the specific resistance decreases with increasing illumination. The lens focuses the image to be televised onto the conductive film.

The external target-electrode ring is connected through a load resistor R_L to a supply of about $+30$ V. The target may be thought of as a large number of elements, each consisting of a small capacitor connected on one side to the target electrode via the transparent conductive film, and shunted by a light-dependent resistor (an element of the photoconductive layer).

When the target is scanned by the beam every part of its surface is brought approximately to cathode potential, establishing a p.d. across the layer. In other words, each tiny capacitor is charged to the potential difference between the target ring and the cathode.

In the dark the photoconductive material is a fairly good insulator, so only a negligible charge escapes before that element is next scanned. However, those target elements which are more or less illuminated will be more or less discharged before the next scan. When the beam next reaches the element and starts to recharge the capacitor a current will flow in the external circuit depending on its state of discharge. Thus a varying current passes through the target load resistor R_L, setting up across it a varying voltage, negative-going for bright elements, which is the 'video' signal and is used to modulate the vision carrier.

The basic principle of the tube known as the plumbicon is the same; it differs in target material, utilizing a far more sensitive lead compound, from which it gets its name.

For reception a cathode-ray tube is used. The fluctuating p.d. from the receiving aerial, after amplification, is applied between the grid of the tube and the cathode (e.g. G and C in Fig. 10.20), and hence the brightness of the spot of light is varied. Two time-bases, connected to scanning coils, cause a scanning motion of the electron beam which must correspond exactly to the motion of the scanning beam at the transmitter. The picture is then reproduced on the screen.

Colour television

The picture tube in a colour receiver is called a 'shadow-mask' tube. This has three separate electron guns, and a screen covered with about $1 \cdot 5 \times 10^5$ phosphorescent dots arranged in 'triads', groups of three, one in each group emitting each of the primary colours red, green and blue. Between the electron guns and the screen is a 'shadow mask', a steel plate having over 5×10^5 holes, each being placed to an accuracy of 1 μm on an area of about 10^3 cm^2. The shadow mask ensures that electrons

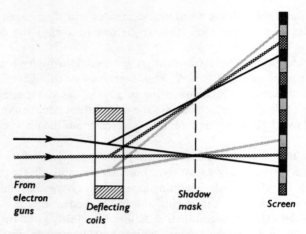

Fig. 10.27. Principle of shadow-mask or tricolour tube.

from one gun can reach only the red dots, and so on. Whereas for a black and white picture, all guns fire with equal intensity and stimulate all dots in a triad so as to give the impression of white light, a colour signal causes the beam intensities to vary so that the primary colours are in the correct ratios to reproduce an impression of the original colour. The 'shadow-mask' principle is shown in Fig. 10.27.

A colour camera is shown in principle in Fig. 10.28. Three separate

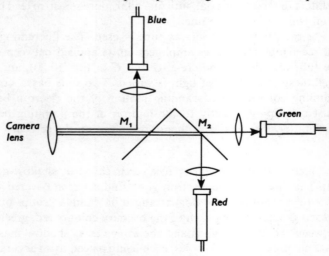

Fig. 10.28. Splitting studio scene into three component colours.

plumbicons provide the red, blue and green signals; the light from the lens system is split into these three colours by dichroic mirrors, the first reflecting blue light but passing red and green, and the second reflecting red and passing green. They are made by depositing extremely thin layers of materials of different refractive indices.

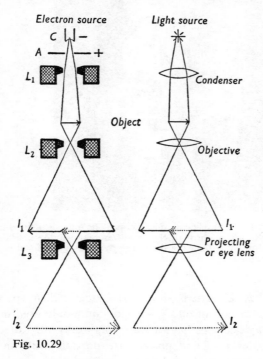

Fig. 10.29

The electron microscope

The resolving power (or ability to show detail) of an optical microscope is limited by the wavelength of light. The image of a point formed by a lens consists of a bright spot surrounded by dark rings. The image systems of two points very close together overlap and it can be shown that two points on the object, which are closer than about half a wavelength, cannot be separately distinguished in the image however great the magnification produced by a microscope.

Electrons may be focused by electric or magnetic fields, and hence they may be used instead of light. Because of the minute size of electrons, an electron microscope might be thought capable of revealing atoms. As far as resolution is concerned, however, electrons behave like

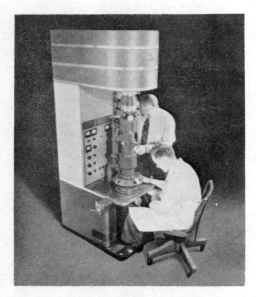

Fig. 10.30. The electron microscope. The man sitting down is observing the image through the small circular windows, while the man standing up is focusing the image by altering the focal length of the projector lens.

waves, and the wavelength is shorter the greater their speed. When accelerated under a p.d. of 60 kV – a p.d. commonly used in electron microscopes – the electrons have a wavelength only 10^{-5} that of light. Thus much greater resolving powers are possible with an electron microscope than can be obtained with an optical one.

The general arrangements of the two instruments are identical, as can be seen from Fig. 10.29. The source of electrons is a heated filament, C, acting as a cathode, and the electrons are given a high speed by an anode, A, maintained at a p.d. with respect to C of about 60 kV. L_1 is a condenser lens which serves to concentrate the electron beam on to the object to be viewed. L_2 is the objective forming an image I_1, and L_3 is the projecting lens throwing a final image I_2 upon a fluorescent screen or a photographic plate. Since electrons are scattered by air the microscope must be evacuated as completely as possible.

A typical form of magnetic lens, used in preference to an electrostatic lens, is shown in Fig. 10.31. It consists of a short solenoid entirely encased in steel, except for a brass ring in the middle of the circular pole

Direction of
electron stream

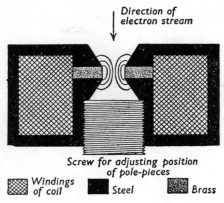

Screw for adjusting position
of pole-pieces

Windings
of coil   Steel   Brass

Fig. 10.31

piece. The magnetic flux would be confined largely to the steel were it not for the brass ring, which causes a leakage of flux into the space in the middle of the pole piece.

Consider an electron moving down through the magnetic lens (Fig. 10.32). There is a component of the flux from left to right at right angles to its direction of motion and applying Fleming's left-hand rule and remembering that the direction of the positive current is opposite to that of the electron, the electron will be found to be urged into the paper. When it has a motion at right angles to the paper, because of the component of the field downwards, it will be urged to the right. An electron nearer to the pole-piece, being in a stronger field, will experience a stronger force to the right. Thus the electrons can be brought to a focus,

Original direction
of electron   Force
perpendicular
to paper

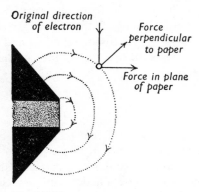

Force in plane
of paper

Fig. 10.32

Fig. 10.33. Microbe of common sore throat (21 500 ×). This has been reduced in size from the original photograph, taken with an electron microscope, which showed a magnification of 43 000 ×.

but at the same time the image is rotated through an angle. (The final image is rotated through an angle of about 60° with respect to the object.

The lenses in the electron microscope are not moved up and down as in the optical microscope; their focal lengths are adjusted by varying the currents through the coils.

Of the electrons which reach the object to be viewed some pass freely while other are stopped by the denser parts. The final image is thus a shadow-graph, similar to that obtained with X-rays (Fig. 10.33). It is necessary that the object should be in the form of a very thin slice or section – a few millionths of an inch thick if the image is to show light and shade. This makes the use of the microscope difficult and restricted although there is very considerable scope for its use, particularly in chemistry, metallurgy and biology.

Summary

Seebeck effect: If the junctions of two different metals in a circuit are kept at different temperatures, a small current flows.

Peltier effect: If a current passes through a circuit containing two different metals, heat is generated at one junction and absorbed at the other.

Thomson effect: If a temperature gradient exists in a conductor an electric potential gradient also exists.

Photoelectric effect: When light falls on certain metals electrons are emitted from them.

$$hf = \phi e + \tfrac{1}{2}mv^2.$$

Thermionic effect: When a metal is heated it emits electrons.

Amplification factor of a triode, $\mu = \dfrac{\delta V_A}{\delta V_G}$.

Questions

1. Give an account of the Seebeck and the Peltier effects showing how you would demonstrate them.

These two effects are said to be reversible effects. What is meant by such a statement? (O)

2. Explain, with diagrams, how you would set up a potentiometer circuit suitable for measuring the e.m.f. of a thermocouple. Indicate the approximate voltages and resistances of the parts of the apparatus if the e.m.f. to be measured is about 0·1 V and the potentiometer wire has a resistance of 5 Ω. (O)

3. A copper-eureka thermocouple has a resistance of 1 Ω and is connected to a millivoltmeter of resistance 10 Ω. When one junction is in ice and the other in steam at 100° C the instrument records 4·17 V. The value calculated from the known constants of the metals is 4·596 mV. How do you account for the discrepancy? (B)

4. Give an account of the Seebeck and Peltier effects, and explain how they are related.

A thermocouple is made up in the usual way, with 2 m of iron wire of 1 mm² cross-section, and 2 m of constantan of the same cross-section. The average resistivity over the range of temperature concerned is 15×10^{-8} Ωm units for iron, and 50×10^{-8} Ωm units for constantan. The e.m.f. of this couple is 0·000 063 V K^{-1} difference in temperature between the two junctions. It is connected in series with a sensitive galvanometer of resistance 5 Ω, and the galvanometer records 5 mA. Find the temperature difference between the junctions of the thermocouple.

5. A thermojunction of iron and copper connected to a galvanometer cannot be used to measure temperatures above about 450 K, whereas one of iron-constantan may be used up to at least 875 K. What is the explanation? (C)

6. Explain in outline the method you would use to verify (a) the law of intermediate metals, (b) the law of intermediate temperatures, as applied to thermocouples. Give full experimental details for one method.

7. An iron-copper thermocouple gives an e.m.f. of 15 μV per degree difference in temperature between hot and cold junctions. What arrangement of six such iron-copper thermocouples, each of resistance 1 Ω, will give the maximum deflection in a galvanometer of resistance $1\tfrac{1}{2}$ Ω? If the galvanometer has a sensitivity

of $\frac{2}{3}$ division per microampere, what is this maximum deflection when the difference in temperature of the junction is 10 K? (CS)

8. Le Chatelier's principle is as follows: 'If one of the factors of any system in equilibrium is changed, thus disturbing the equilibrium, the effect produced tends to restore that factor to its original value.'

Discuss the application of the principle to the phenomena of thermoelectricity, the thermionic effect and self-induction.

9. Describe a form of photoelectric cell and give a short account of the uses to which such a cell can be put. (CS)

10. The following facts are known about emission of photoelectrons from a metal surface: (a) the rate of emission is proportional to the intensity of incident light, (b) the ratio emission/intensity varies with the wave-length, (c) the maximum velocity of emission is given by

$$v^2 \propto \left(\frac{1}{\lambda} - \frac{1}{\lambda_0} \right),$$

where λ is the wave-length and λ_0 a constant characteristic of the metal. Devise a means of checking these facts. (OS)

11. Explain the construction and mode of action of a *diode valve*. If the filament current is kept constant, how will the current through the valve vary when the plate voltage is gradually increased? What would occur if the experiment were repeated with a slightly higher filament current? (B)

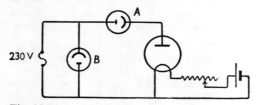

Fig. 10.34

12. A neon lamp has two electrodes. When connected to d.c. the positive electrode glows, and when connected to a.c. both electrodes glow. Explain which electrodes glow in the neon lamps A and B in the circuit of Fig. 10.34.

13. How may the diode be employed to rectify an alternating current? (B)

14. A triode consists of a filament, grid and anode in an evacuated bulb. Regarding the filament as a plane metallic sheet emitting electrons with negligible velocities, the grid as a plane mesh parallel to the filament and 2 mm from it, the plate as a plane solid metallic sheet 12 mm from the grid, and assuming that uniform electric fields exist between filament and grid and between grid and plate, what must be the relative potentials of the grid and plate with respect to the filament for electrons to reach the plate? What important factor do you ignore?

15. Describe a cathode-ray oscillograph and given an account of some of its applications. (OS)

16. If 100 V d.c. are applied between the Y-plates of a cathode-ray tube, the spot is deflected vertically 1 cm. A sinusoidal alternating voltage is applied to the same plates, and a suitable time-scale to the X-plates, so as to display the form of the wave. If the vertical distance between the crest and the trough of the sine wave is 4 cm, how much power will be dissipated when the same voltage is applied to a resistance of 100 Ω ?   (CS)

17. Write a short account of the electron microscope.

18. Electrons from a heated filament are accelerated by a potential difference of 200 V and then pass horizontally as a fine pencil beam mid-way between two parallel deflecting plates 2 cm long and 0·6 cm apart. The electrons are received and form a spot on a flat screen 15 cm from the near end of the deflecting plates. Show that the velocity of the electrons just before they reach the deflecting plates is $8·39 \times 10^6$ m s^{-1}. Calculate the deflection of the spot formed on the screen when a potential difference of 30 V is applied between the deflecting plates. Calculate also the magnitude of the magnetic induction which, applied perpendicular to both the path of the electron beam and the electric field between the deflecting plates, would restore the spot to its original position on the screen. State the assumptions and approximations used in each of the three calculations.

(e/m for the electron $= 1·76 \times 10^{11}$ C kg^{-1}.)   (O & C 1971)

19. Write down Einstein's equation for photoelectric emission. Explain the meaning of the terms in the equation and discuss its significance.

Describe briefly how Einstein's equation may be verified experimentally.

An effective point source emits monochromatic light of wavelength 4500 Å at a rate of 0·11 W. How many photons leave the source per second? Light from the source is emitted uniformly in all directions and falls normally on the cathode of area π cm^2 of a photocell at a distance of 50 cm from the source. Calculate the photoelectric current, assuming 10% of the photons incident on the cathode liberate electrons.

(Planck's constant $= 6·6 \times 10^{-34}$ J s^{-1}; charge of electron $= 1·6 \times 10^{-19}$ C; 10 Å $= 1$ nm.)   (O & C 1968)

20. Give a brief description of the construction of a high-vacuum diode. Draw a graph which shows the variation of the current through such a diode with the potential difference across it, and account for the main features of the curve.

Describe how the introduction of a third electrode, the grid, makes possible (a) control of the current which reaches the anode, and (b) the amplification of a voltage.   (O & C 1968)

21. How would you attempt to measure the maximum velocity with which photoelectrons leave a metal surface irradiated by monochromatic light?

Explain how this velocity depends on the wavelength of the light, the intensity of the light and the nature of the metal.

When the incident light is monochromatic and of wave-length 600 nm, the kinetic energy of the fastest electrons is $2·56 \times 10^{-20}$ J. For light of wavelength 400 nm, it is $1·92 \times 10^{-19}$ J. Use this information to estimate the value of Planck's constant h and the work function of the metal.

(Take the velocity of light to be 3×10^8 m s^{-1}.)   (O 1971)

22. Draw a clear labelled diagram of a cathode-ray oscilloscope. Include on this diagram the control grid (brilliance control) and the focusing anode, and indicate typical voltages applied to the various electrodes.

Explain how the oscilloscope can be used as an instrument to measure both direct and alternating voltages.

The accelerating voltage applied in an oscilloscope is 3000 V, and the Y-deflection plates are 3 mm apart and 4 cm long. Calculate the vertical displacement of the beam after passing between these plates when the p.d. between them is 20 V.   (O 1968)

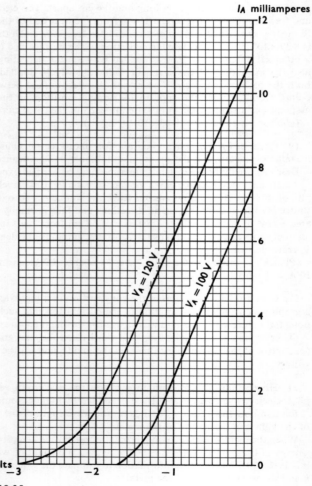

Fig. 10.35

23. Give an account of the chief features of thermionic emission. Describe the thermionic triode and explain how you would obtain experimentally, for a triode, a set of mutual characteristics such as those shown in Fig. 10.35. Use this figure to obtain, for the triode concerned, values of (*a*) the amplification factor μ, (*b*) the mutual conductance g, (*c*) the impedance r.

Draw a circuit diagram to show a single triode acting as an amplifier.

(O 1968)

11 Radio transmission and reception

In 1865 Clerk Maxwell put forward the theory that electromagnetic waves should exist in the aether, that they should be generated by electric oscillations and that they should travel with the velocity of light.

The experiments of Hertz

In 1888 Heinrich Hertz produced electromagnetic waves. His oscillator (Fig. 11.1) consisted of an induction coil, to the secondary terminals of which were connected two brass plates 40 cm square joined by copper wires 30 cm in length to two balls forming a spark gap 2–3 cm wide.

His receiver was a loop of wire of about 35 cm radius with a small, adjustable, spark gap. When placed near the oscillator, the waves from the latter induced currents in it and small sparks jumped across the gap.

Hertz showed that the velocity of the waves was equal to that of light and that, like light, they underwent reflection, refraction, interference, and polarization. 'Hertz's researches were one of the most marvellous triumphs of experimental skill, of ingenuity, of caution in drawing conclusions, in the whole history of physics', wrote Sir J. J. Thomson. 'Younger physicists, with the very efficient means of detecting electrical waves now at their command will not realize the difficulties of these

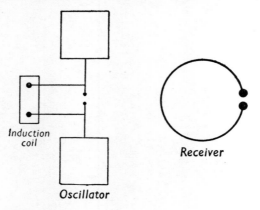

Induction coil

Oscillator

Receiver

Fig. 11.1

JAMES CLERK MAXWELL   HEINRICH HERTZ

JAMES CLERK MAXWELL (1831–1879) *is best remembered for his famous equations which summarized electromagnetic theory, and predicted the existence of electromagnetic waves. His ideas were based on his concept of the aether, 'that supreme paradox of Victorian science and yet a triumph of the scientific imagination'. At the age of sixteen he went to Edinburgh University and three years later to Cambridge. A fellowship at Trinity College, Cambridge, and professorships at Aberdeen, at King's College, London, and then at Cambridge form the main events of his career. He left his mark at Cambridge, quite apart from his scientific discoveries, in the erection of the Cavendish Laboratory. His main contributions to physics were his electromagnetic theory of light, the mathematical development of the kinetic theory of gases and his work in colour vision. He had an impish streak in his character and delighted, to the end of his days, in writing humorous verse on scientific topics.*

HEINRICH HERTZ (1857–1894) *was the first to establish the existence of electromagnetic waves. Maxwell's theory had shown that such waves should be generated by electrical oscillations, and that their velocity should be equal to that of light. It was realized that the oscillations must be very rapid, otherwise the wavelength would be of the order of miles, and that very rapid oscillations were produced by the spark discharge from capacitors. Several investigators were trying to detect electromagnetic waves, including Sir Oliver Lodge, who said later, 'Hertz stepped in before the English physicists, and brilliantly carried off the prize'. Hertz's great achievement brought him the professorship at Bonn at the early age of thirty-two, but his life was cut short by blood poisoning only five years later.*

experiments, but older ones, who like myself, began by using Hertz's method, and had to observe whether tiny sparks only a fraction of a millimetre long waxed or waned when the detector was moved from one position to another, will remember how arduous and harassing these experiments were and how long it took to make sure that the effects observed were not spurious.'

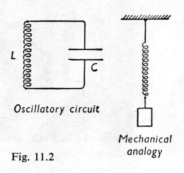

Oscillatory circuit

Mechanical analogy

Fig. 11.2

The oscillatory circuit

The circuit of Hertz's oscillator is essentially the same as Fig. 11.2, which is called an oscillatory circuit. The two brass plates in Hertz's oscillator act as the plates of a capacitor. Each spark that passes across the spark gap sets up a train of very high frequency electrical oscillations in the circuit. Each spark itself is therefore an oscillatory discharge.

To produce electrical oscillations, a circuit must possess inductance, L, and capacitance, C. The frequency of the oscillations, f, is given by

$$f = \frac{1}{2\pi\sqrt{LC}} \quad \text{(see p. 213)}.$$

If the inductance is expressed in H and the capacitance in F, the frequency is given in Hz.

It is, perhaps, helpful to compare electrical oscillations with the mechanical oscillations of a mass on the end of a spiral spring. The alternate charging of the capacitor in opposite directions is comparable to the stretching and compressing of the spiral spring; the inertia of the electric current due to the inductance is comparable to the mass. Each discharge of the capacitor builds up round the inductance a magnetic field which then collapses. While the current and the magnetic field are increasing, the e.m.f. due to the inductance opposes the current; but while the current is decreasing and the field is collapsing, the induced

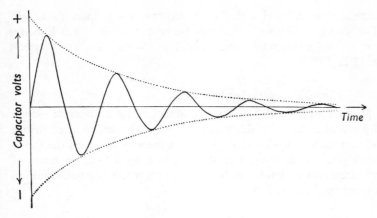

Fig. 11.3

e.m.f. is in the same direction as the current and so causes the current to be prolonged and the capacitor to be charged in the opposite direction. Similarly the mass, when the spiral spring is fully extended, opposes the shortening of the spring at first, but when the natural length has been reached, the kinetic energy causes the upward motion to continue so that the spring is compressed.

The mechanical oscillations gradually die away owing to loss of energy through air resistance; similarly, the electrical oscillations decrease owing to loss of energy in the form of heat in the resistance of the circuit. A train of damped oscillations is shown in Figs. 11.3 and 11.4. The frequency of the exciting sparks is usually sufficiently low to ensure that one train of waves dies away before the next is initiated.

Hertz's circuit differs from that in Fig. 11.2 in that the plates of his

Fig. 11.4. Trace showing the oscillatory discharge of a capacitor, taken with a cathode-ray oscillograph. Note how the brightness of the trace varies with the writing speed.

capacitor are not parallel and close together but are opened out. His circuit is called an 'open' oscillatory circuit and is a much better radiator of electromagnetic waves than the 'closed' oscillatory circuit of Fig. 11.2.

The mechanism of radiation

Many people find it helpful to devise a model or some concrete picture of a physical phenomenon; this is perhaps more characteristic of British than of continental physicists. A picture of the way in which the radiation takes place from Hertz's oscillator has been devised in terms of tubes of electric flux. A tube of flux may be regarded as a collection of, or bounded by, lines of flux.

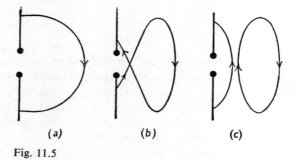

(a)   (b)   (c)

Fig. 11.5

Suppose the upper half of the spark gap in Hertz's oscillator is charged positively and the lower half is charged negatively. Tubes of flux, starting on a positive charge and ending on an equal negative charge, will lie around the spark gap, like the tube shown in Fig. 11.5 (a). When the rapid discharge across the spark gap commences, the ends of the tubes of flux will be jerked across the gap and a loop will be formed rather like the loop in a skipping rope when its ends are jerked across each other. This loop breaks away; it is no longer anchored to positive and negative charges which must be dragged along with it and it moves outwards at right angles to itself, with the velocity of light, continually expanding as it goes. It creates a magnetic field perpendicular both to itself and to its direction of motion.

Meanwhile the other two parts of the tube have joined together and stretch across the spark gap, now charged in the opposite direction. The process is repeated and, since the discharge oscillates backwards and forwards, a succession of oppositely directed loops will be radiated

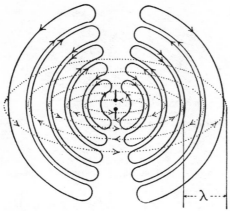

Fig. 11.6

as in Fig. 11.6. The dotted lines represent the lines of magnetic flux (in the horizontal plane only).

Only a single tube has been shown in each loop, but actually there should be a number of tubes. The distribution of tubes and magnetic lines of flux is shown in Figs. 11.7 and 11.8 which represent the fields along a straight line from the middle of the spark gap. The fields vary sinusoidally and hence the tubes and magnetic lines of flux are spaced as shown. The distance of any point on the sine curves from the axis represents the strength of the field.

In the wave the electric and magnetic fields are in phase; they reach their maximum and zero values simultaneously. At the oscillator, however, there is a phase difference of 90° between the electric and magnetic fields. The magnetic field at the oscillator is a maximum when the current across the spark gap is a maximum and thus when the electric

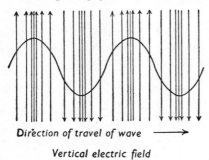

Direction of travel of wave ⟶

Vertical electric field

Fig. 11.7

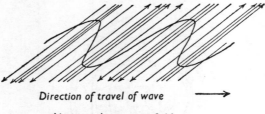

Direction of travel of wave ⟶

Horizontal magnetic field

Fig. 11.8

field is zero. The change of phase may be considered to occur as a result of the breaking away of a loop.

On arrival at the receiver a loop must break and its free ends must become attached to positive and negative charges. The movement of the negative charge as the tube collapses constitutes an electric current in the receiver.

Wavelength and frequency

The distance between similarly directed loops of electric field lines, or magnetic field lines in a wave is called the wavelength, λ (Fig. 11.6). If f waves are radiated per second and v is the velocity of the waves, it is clear that $v = f\lambda$.

Now $v = 3 \times 10^8$ m s^{-1}, and in Hertz's experiment $f = 5 \cdot 5 \times 10^7$ Hz, hence

$$\lambda = \frac{3 \times 10^8}{5 \cdot 5 \times 10^7} = 5 \cdot 4 \text{ m}.$$

The frequency used in radio varies from about 100 000 Hz, i.e. 100 kHz, to 3×10^{10} Hz, i.e. 30 000 MHz; the corresponding wavelengths are 3 000 m and 1 cm.

To radiate appreciable energy it is essential that the frequency should be high. The alternating current in the National Grid, for example, with a frequency of 50 Hz, radiates negligible energy.

Marconi

The commercial application of electromagnetic waves received its chief and primary impetus from Guglielmo Marconi (1874–1937). Although he received a Nobel prize and more honours than any other man of his time, Marconi's gifts were as much those of a good business man as of a physicist and inventor.

Marconi was the son of an Irish mother and an Italian father and hence he spoke English as well as Italian. He made his first experiments in the gardens of his home, the Villa Griffone, near Bologna, with his father's gardeners as assistants. 'The idea of transmitting messages through space by means of etheric waves', he said, 'came to me suddenly as a result of having read in an Italian electrical journal about the work and experiments of Hertz. . . . My chief trouble was that the idea was so elementary, so simple in logic, that it seemed difficult to me to believe no one else had thought of putting it into practice. Surely, I argued, there must be much more mature scientists than myself, who had followed the same line of thought and arrived at an almost similar conclusion.'

In 1896, after he had succeeded in transmitting messages a distance of 2800 m, he came to England where his mother had relatives and friends. The Engineer-in-Chief of the G.P.O. became interested in his experiments and a test transmission took place from the roof of the post office at St Martins-le-Grand to a receiving station on the Thames Embankment. Then followed a series of transmissions over increasing distances, $6\frac{1}{2}$, $10\frac{1}{2}$, 13 km over Salisbury Plain, over the Bristol Channel from Penarth to Bream Down, across the English Channel from Wimereux to South Foreland, from St Catherine's Point, Isle of Wight to the Lizard, Cornwall, and finally the famous transmission across the Atlantic.

It was thought that it would be impossible to send signals across the Atlantic since, owing to the curvature of the earth, the waves would pass out into space. Nevertheless, Marconi determined to try.

He erected a powerful transmitting station at Poldhu, Cornwall. He was, at this time, still using a spark transmitter similar in principle to that of Hertz, but he had discovered the value of a high aerial. His receiver was much more sensitive than that of Hertz, its main component being a 'coherer', consisting of filings of nickel and silver between two silver cylinders; these filings cohered when electrical oscillations from the receiving aerial passed through them, and the resulting decrease in resistance enabled a battery to supply a current and cause a click in telephones or a mark on a moving tape.

Marconi sailed for St John's, Newfoundland, and set up his receiving station in the old military barracks on Signal Hill. For nearly a fortnight the winds were tremendous. Balloons sent up to raise the receiving aerial broke away and disappeared in heavy mist. On 12 December 1901 he managed to fly a kite despite the gale and drenching rain. About 1230 pm three faint clicks were heard in his receiver, and the letter S, the prearranged signal from Poldhu, 2880 km away, appeared on his tape machine.

The suggestion that Marconi might have picked up stray signals proved untrue. In a short time a station was set up at Glace Bay on Cape Breton coast in Canada, and transmission across the Atlantic was in regular operation.

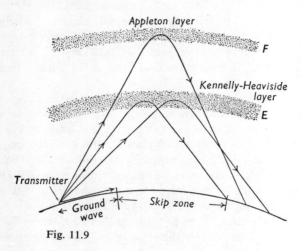

Fig. 11.9

The ionosphere

In 1902 Kennelly and Heaviside independently put forward an explanation of Marconi's success in transmitting waves across the Atlantic, despite the curvature of the earth. They suggested that there was an electrical reflector in the upper atmosphere, now called the Kennelly–Heaviside layer. In 1924–5 Dr (later Sir Edward) Appleton proved its existence by experiment and found its height to be about 100 km above the earth. Later Appleton and his collaborators fired short pulses of electromagnetic waves at the layer – a method now used for radar; they found that some of the pulses were getting through and being reflected from a height of about 240–290 km above the earth. In this way the Appleton layer was discovered (Fig. 11.9).

The upper atmosphere is known to be ionized and hence is called the ionosphere; the rarefied air contains many charged molecules due to electrons ejected from the sun. Research on the ionosphere, now proceeding all over the world, has shown that there are several, distinct, conducting layers. The Kennelly–Heaviside layer is called the *E* region,

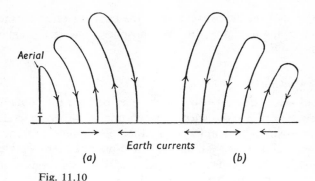

Fig. 11.10

and the Appleton layer the F region. The Kennelly–Heaviside layer rises during the night and drops at sunrise; its reflecting properties are more regular at night.

Electromagnetic waves penetrate a distance into the Kennelly–Heaviside layer depending on their wavelength and the obliquity of their incidence; the shorter the wave length, the greater the penetration and the greater the obliquity the less the penetration. Short waves, unless they are very oblique, go right through. Since the Appleton layer is about four times as strong a reflector as the Kennelly–Heaviside layer, owing to the greater concentration of ions, short waves are usually reflected there.

The electromagnetic waves from a transmitting aerial can be imagined to fall into two parts; that part which travels towards the sky may be refracted in the ionosphere and returned towards earth, and that part which sets off along the ground continues along the ground.

If an aerial is earthed and electrical oscillations are fed in at the bottom, an arrangement due to Marconi, waves are sent out as in Fig. 11.10 (a). The lower ends of the lines of flux travel along the earth and give rise to earth currents. Since the earth is not a perfect conductor the earth currents travel more slowly than the field lines above and hence the waves are bent downwards as in Fig. 11.10 (b).

The range of the ground waves depends on the conductivity of the soil. Transmission is poor over sandy soil, and best over sea water. There is often a region, called the skip zone (Fig. 11.9), where no waves are received. This region is too far away from the transmitter for the ground waves to reach it, but not sufficiently far away to receive the nearest sky waves reflected from the ionosphere.

Fading, in wireless reception, is due to interference between the ground waves and the sky waves; when the path traversed by the sky waves alters, owing to changes in the ionosphere, interference maxima and minima result.

Modulation

The damped waves generated by sparks may be used for communication by morse; but for radio telephony continuous waves, generated by valves, or transistors, are employed and their amplitudes, or frequency, is varied by sound waves falling on a microphone, a process called *modulation* (Fig. 11.11). The frequencies of the radio waves in the broadcast medium wave-band are of the order of 10^6 Hz, and the frequencies of normal sound waves are of the order of 10^2 or 10^3 Hz; thus Fig. 11.11 is not drawn to scale.

Modulated waves are sometimes said to have a radio-frequency (r.f.)

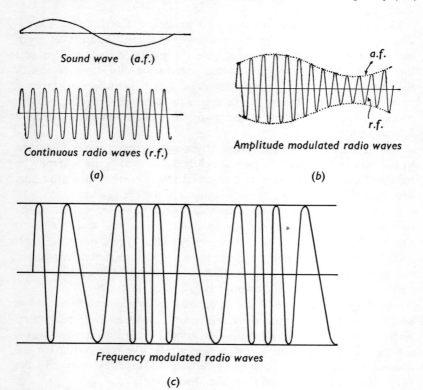

Sound wave   (a.f.)

Continuous radio waves (r.f.)

(a)

Amplitude modulated radio waves

(b)

Frequency modulated radio waves

(c)

Fig. 11.11

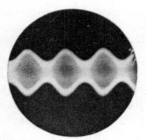

Fig. 11.12. Trace, taken with a cathode-ray oscillograph, showing a carrier wave of frequency 85 kHz amplitude modulated to a depth of about 50 per cent with a note of frequency 400 Hz.

component, and an audio-frequency (a.f.) component. It is, of course, the a.f. that must be reproduced by the receiver, and the waves of r.f. are merely carrier waves.*

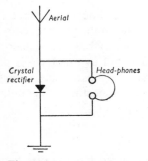

Fig. 11.13

Detection

Figure 11.13 represents a very simple receiver which would function under certain circumstances. The waves from the transmitter set up r.f. currents, whose amplitudes vary at a.f., between the aerial and the earth. It may be asked why the circuit cannot be made still simpler by omitting the crystal rectifier and putting the headphones between aerial and earth. The currents between aerial and earth are of r.f., and the telephone diaphragm cannot vibrate at so high a frequency; like the coil of a moving-coil galvanometer connected to an ac supply, the diaphragm

* In another form of modulation, known as frequency modulation, the amplitude of the carrier waves is kept constant and their frequency is varied at the frequency of the audio waves, the extent of the variation in frequency being proportional to the amplitude of the audio waves (Fig. 11.11 (c)).

would be urged first in one direction, then in the other, so rapidly that it remained at rest.

Clearly some method of separating, as it were, the a.f. from the r.f. is necessary. A crystal of galena (i.e. sulphide of lead) touched with a phosphor-bronze wire called a 'cat's-whisker', has the property of allowing an electric current to pass through it much more readily in one direction than in the other. Thus the current through the crystal between aerial and earth is as shown in Fig. 11.14; the bottom halves of the carrier waves are suppressed. The half-carrier waves passing through

Detection

Fig. 11.14

the crystal are all in the same direction and hence reinforce each other. The curve of their average value is similar to the a.f. frequency wave, and hence the diaphragms of the headphones vibrate in synchronism with the a.f. The crystal acts as a rectifier, but as this term is usually employed with reference to the conversion of a.c. to d.c. in power circuits, the diode is more commonly called a detector or demodulator; the process is called detection or demodulation.

Tuning

The simple receiving circuit just considered has the disadvantage that it cannot be tuned. The circuit of aerial and earth is essentially the same as the oscillatory circuit of Fig. 11.2 and contains inductance, L, and capacitance, C, a capacitor being formed by the aerial and earth. Its resonant frequency is $1/\{2\pi\surd(LC)\}$ (see p. 213). Only if its resonant frequency corresponds to the frequency of some broadcasting station will it reproduce signals.

Tuning can be achieved by means of a variable capacitor, or a variable inductance or both (Fig. 11.15). The capacitor C_1 is in series with the capacitor formed by the aerial and the earth and hence reduces the total capacitance, thus increasing the resonant frequency of the circuit. The inductance L increases the total inductance of the circuit and hence reduces the resonant frequency of the circuit.

The crystal has a high internal resistance and if placed between aerial and earth as in Fig. 11.13 it would make the tuning of the circuit flat. Its

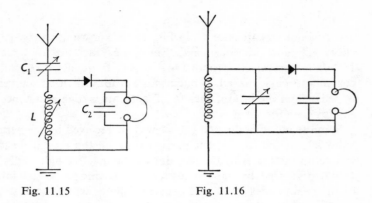

Fig. 11.15   Fig. 11.16

change of position in Fig. 11.15 is therefore an improvement. The function of the capacitor C_2 will be considered later.

The tuning circuit just described is an acceptor circuit (see p. 214).
When tuned it has a minimum impedance and takes a maximum current; thus there is a maximum p.d. across the inductance, to be passed
on to the headphones. An alternative tuning circuit is the rejector circuit
shown in Fig. 11.16, which has a very high impedance at the resonant
frequency and also gives a high p.d. across the inductance.

The examples so far given are called single circuit tuning and are not
very selective, i.e. they do not permit the separate reception of signals
whose frequencies are near to one another. By coupling two circuits
together and tuning both of them, greater selectivity is obtained. In
Fig. 11.17 both circuits are acceptor circuits since the applied e.m.f. is in
series with the inductance and capacitor in each case. The coupling
takes place through the mutual inductance of the two coils.

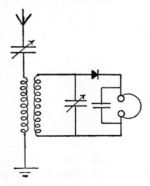

Fig. 11.17

The by-pass capacitor

The function of capacitor C_2 in Fig. 11.15, known as the by-pass capacitor, will now be considered. Since the reactance of a capacitor, $1/(2\pi fC)$, is inversely proportional to the frequency f, C_2 allows r.f. currents to pass through it much more readily than a.f. currents. The impedance of the headphones to r.f. currents is high and hence C_2 acts as a by-pass for them.

Suppose an unmodulated carrier wave is received by the aerial. Each half-cycle of the r.f. current passing through the crystal detector will charge up the plates of C_2 and, during the inactive half-cycles, C_2 will tend to discharge through the headphones. Owing to the high impedance of the headphones at r.f. the capacitor plates will remain at a steady voltage, nearly the peak value, and the diaphragm will not vibrate because the current through the headphones is steady. If, however, a modulated carrier wave is being received, the p.d. across C_2 will vary at a.f. and hence a current varying at a.f. will pass through the headphones. The capacitance of C_2 is kept small so as to avoid the possibility of a smoothing out of the a.f. signal.

The diode as a detector

A diode allows current to pass through it in one direction only, as was explained on p. 231, and hence a thermionic diode can be used instead of a semiconductor diode as a detector (Fig. 11.18).

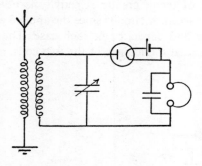

Fig. 11.18

After a period of disuse the semiconductor diode became the most popular detector because it introduces very little distortion and it is small and robust, and cheaper.

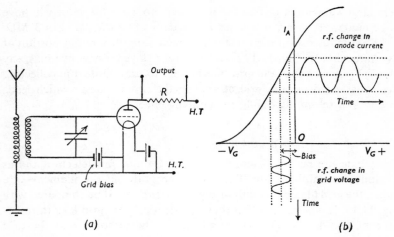

Fig. 11.19

The triode as an amplifier

The meaning of the amplification factor, μ, essential to the understanding of the following account of the use of the triode as an amplifier, was described on p. 236.

The grid and filament of the triode are connected to the ends of the tuning circuit as in Fig. 11.19, and hence there is a p.d. between them varying at r.f. The anode is connected to an 'anode load', which may take the form of a high resistance, R, across which amplified p.d.'s are produced. This is connected to the next stage in the receiver, possibly a detector.

It was shown on p. 236 that a change in the grid voltage of V_A has the same effect on the filament-anode current as a change of μV_A on the anode. But in Fig. 11.19 the current must also be changed in the anode load, R.

Suppose the impedance of the valve to a.c., defined as

$$\frac{\text{change in anode potential}}{\text{corresponding change in anode current}}$$

is R_a (typical values vary from 5000 to 30000 Ω). The voltage change of μV_G will be across the total resistance of the anode circuit, $R + R_a$. Hence the voltage change across the anode load is $\{R/(R+R_a)\}\mu V_G$. The expression $\mu R/(R+R_a)$ is called the *voltage amplification factor*. The larger the value of R the nearer this approximates to μ. If, however, a very large value of R is used the HT voltage to work the valve will have to be made very large, and hence R is usually not more than 2 or 3 MΩ.

To avoid distortion the valve must operate on the straight portion of its characteristic (Fig. 11.19 (b)). The grid is given a negative bias, say of -2 V, by means of the grid-bias battery to prevent it from collecting electrons and causing a grid current in the tuning circuit which leads to a loss of selectivity due to its damping effect.

The triode as a detector

The triode can be used simultaneously as a diode for detection and as a triode for amplification. Regarding the grid and the filament as a diode, the grid filament circuit of Fig. 11.20 (a) should be compared with Fig. 11.18. It will be seen that they are identical; the grid leak (a resistance of about 1 MΩ) corresponds to the headphones and the grid capacitor corresponds to the by-pass capacitor.

Suppose that a constant r.f. signal is applied to the circuit in Fig. 11.20 (a). The grid capacitor is no barrier to r.f. currents and during the half-cycles when the signal causes a positive potential to be applied to the grid, the latter collects electrons. These electrons leak away slowly through the grid leak and the coil of the tuning circuit back to the filament. After several r.f. cycles an equilibrium is attained between the collection of electrons and their escape through the grid leak, the grid acquiring a negative or less positive potential and causing a decrease in the anode current.

When the strength of the r.f. is varied, the grid voltage and the anode current vary in the same manner. Hence a signal modulated at audiofrequency gives rise to an a.f. variation in the anode current.

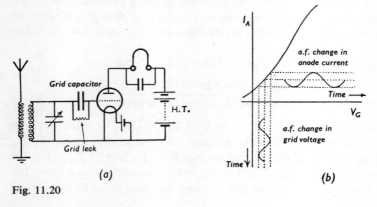

Fig. 11.20

Figure 11.20 (*b*) shows the effect of an a.f. wave on the voltage of the grid and the corresponding change in the anode current.

Clearly the grid leak could equally well have been connected directly between grid and filament.

The triode should be operated on the straight part of the characteristic with a zero or slightly positive grid bias since, in contrast to the use of a triode as an amplifier only, a grid current is required. This method of using the triode as a detector is known as *grid rectification*.

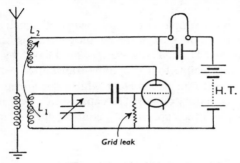

Fig. 11.21

Reaction

The strength of the signal reproduced by the receiver can be increased by a method variously known as reaction, regeneration, feedback, or back coupling.

Figure 11.21 represents a circuit consisting of a single triode used as a grid detector, and incorporating a 'tickler' coil L_2 to provide reaction. L_2 is connected in the anode circuit and is coupled with one of the tuning coils L_1. The alternating currents in L_1 and L_2 are in phase and hence the currents in L_2 will induce an increase in those in L_1, thereby increasing the currents through the valve and hence in L_2 itself. If L_1 and L_2 are coupled too closely the valve will behave as an oscillator and 'howling' will be heard in the headphones. The amount of feedback, as it is called, can be controlled by varying the relative positions of L_1 and L_2 or by inserting a variable capacitor (not shown in Fig. 11.21) between the anode and filament of the valve, which acts as a by-pass and enables the current in L_2 to be varied.

The valve as an oscillator

Energy from the anode circuit is fed back into the grid circuit to produce reaction. If this energy compensates for the energy lost as heat in

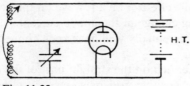

Fig. 11.22

the grid circuit, continuous oscillations will result and the valve will behave as an oscillator (Fig. 11.22).

Figure 11.23 shows a simple circuit by means of which modulated waves can be generated. The part of the circuit generating carrier waves is similar to that of Fig. 11.22. The waves are modulated by the microphone M, similar in principle to a telephone transmitter. Sound waves falling on M give rise to an a.f. current which is coupled through a transformer to the oscillatory circuit. The capacitor C_1 acts as a by-pass for the r.f. currents, and the capacitor C_2 controls their frequency.

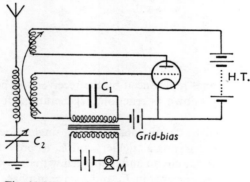

Fig. 11.23

Coupling of valves

In a simple receiver there may be three stages whose functions are (1) r.f. or high-frequency amplification, (2) detection, (3) a.f. or low-frequency amplification. One or more valves may be used in each stage and hence it is necessary to consider how valves can be coupled.

Coupling can be effected by four chief methods, resistance-capacity, choke-capacity, tuned anode and transformer coupling.

Figure 11.24 (a) illustrates resistance-capacity coupling. As has already been explained, the p.d. across R is $\{R/(R+R_a)\}\mu V_G$ and hence varies with V_G. The upper end of R is maintained at the positive potential of

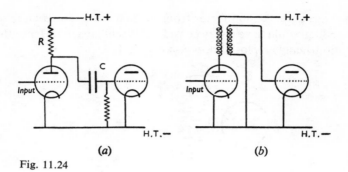

(a)   (b)

Fig. 11.24

the HT supply and the potential of the lower end varies. If the lower end of R were connected direct to the grid of the second valve, the grid would receive current from the HT supply, and behave as an anode. Hence a capacitor C, which allows alternating current to flow but not direct current, is inserted. In order that C shall not insulate the grid completely a grid leak is provided and this enables electrons collected by the grid to leak back to the filament.

The resistance R may be replaced by a choke, the rest of the circuit remaining unchanged; this is choke-capacity coupling. The advantage of a choke is that while it has a high impedance for a.c. its resistance for d.c. is low, and it permits moderate anode voltages to be used. It has the disadvantages that its impedance varies with frequency and that it has considerable self-capacity. In consequence, low notes and very high ones are not well amplified.

The resistance R may be replaced also by a rejector circuit, consisting of an inductance and variable capacitance, the remainder of the circuit being unchanged; this is tuned anode coupling. As explained on p. 214 a rejector circuit has a high impedance at its resonant frequency. As this is a tuned system it is applicable to one frequency only and it increases selectivity. It cannot be used in the audio-frequency stage because the amplification there must be independent of frequency.

Figure 11.24 (b) illustrates transformer coupling. A capacitor and a grid leak are unnecessary here because there is no direct connection between the anode of the first valve and the grid of the second. Air-cored transformers are used for r.f. coupling and iron-cored ones for a.f. For r.f. the secondary of the transformer is sometimes tuned by connecting across it a variable capacitor. It might seem possible greatly to increase the amplification by using a step-up transformer of high ratio. However, the inductance of the primary must be high in order to give a large voltage amplification factor from the first valve, and a large secondary winding would have a very considerable self-capacity. Hence the step-up ratio usually employed does not exceed 1:6.

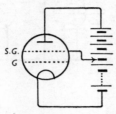

Fig. 11.25

Tetrode and pentode

The triode, particularly when used as a r.f. amplifier, is liable to be unstable and to oscillate. The anode and the grid act as the plates of a capacitor and energy is fed back from the anode circuit to the grid circuit through the capacitance. The defect can be overcome by inserting a screening grid between the anode and the control grid (Fig. 11.25). The resulting valve is called a tetrode or a screened-grid valve. If the potential of the anode is 120 V that of the screening grid is usually about 80 V. High voltage amplification factors, from 100 to 1500, can be obtained with a tetrode.

When the electrons from the filament strike the anode with sufficient velocity they cause the emission of secondary electrons, and if the potential of the screening grid happens to be higher than that of the anode, it will collect these secondary electrons. The kink in Fig. 11.26 shows how, after secondary electrons begin to be emitted, the anode current of a tetrode actually decreases with increasing anode potential for a certain range of voltage. Thus distortion will result if the potential of the anode swings below that of the screening grid. The insertion

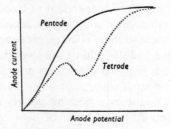

Fig. 11.26

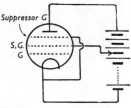

Fig. 11.27

between the anode and the screening grid of a suppressor grid, kept at earth potential, enables the anode to attract back the secondary electrons without losing any to the screening grid. The resulting five-electrode valve (Fig. 11.27), called a pentode, is often used in preference to either a tetrode or a triode for amplification purposes.

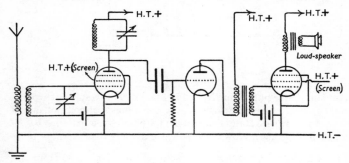

Fig. 11.28

Three-valve receiver

Figure 11.28 represents a simple three-valve receiver. A pentode r.f. amplifier is coupled by a tuned anode circuit to a triode acting as a grid detector. The triode is coupled through a transformer to a pentode a.f. amplifier. The loud speaker is of the low-impedance type, and hence is connected through a step-down transformer to the anode circuit of the last valve.

Questions

1. Explain what is meant by an *oscillatory electric circuit* and distinguish between *closed* and *open* circuits. Describe how oscillatory circuits are used in wireless transmission and reception.   (N)

2. Describe the triode thermionic valve. Give an account of two applications of a thermionic valve, with a brief explanation of the action in each case.   (N)

3. Discuss the physical principles involved in the use of a triode in the reception of wireless signals as (*a*) a rectifier, (*b*) an amplifier.   (N)

4. Describe the triode valve and give a short account of its mode of action. Explain the term *voltage amplification* of the valve and describe how you would determine it experimentally. Give a labelled diagram of the circuit used.   (N)

5. Describe briefly the action of a modern radio receiver.   (OS)

6. It has been reported in *The Times* that members of the Australian forces cut off and left behind in Timor made a wireless set and communicated with Australia. Enumerate the components which you consider they could improvise from derelict cars and machinery and explain how you would set about the task if you found yourself in their position.

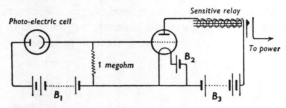

Fig. 11.29

7. When light falls on the photoemissive cell in Fig. 11.29 the resulting current is amplified and then operates a relay which switches on a power circuit not shown. Explain how the circuit works, stating clearly the purpose of the batteries B_1, B_2, B_3, and the megohm resistance. How does the valve amplify the current?

8. Explain what is meant by reaction in a radio receiver. Fig. 11.30 represents 'capacitive reaction'. Explain carefully how it works.

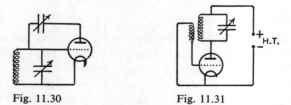

Fig. 11.30   Fig. 11.31

9. The circuit in Fig. 11.31 represents a valve oscillator. Explain how it works and how it differs from Fig. 11.22.

10. Draw the circuit of a receiver consisting of three triodes: a h.f. amplifier, coupled by a tuned anode circuit to a grid rectifier, coupled by resistance-capacity to a l.f. amplifier.

11. Draw the circuit of a two-valve receiver consisting of a detector coupled by a transformer to a l.f. amplifier. Include reaction.

12. The product of *wavelength* and *frequency* of *electromagnetic radiation* in a medium is equal to the *velocity of the radiation*. Explain the meaning of the italicized terms in this statement.

Which of the following quantities change when electromagnetic radiation goes from one medium to another:

frequency, wavelength, velocity?

Give what observational and/or theoretical justification you can to support your statements.   (O & C 1970)

13. Write an essay on the generation and detection of radio waves. Divide your essay into three main sections:

1. How oscillations are maintained in an LC circuit, and how they are modulated by a signal.

2. How the oscillations are converted into radio waves, and how these are received to produce oscillations in a tuned circuit.

3. How, by rectification, the signal which modulates the carrier wave can be detected.   (O 1969)

12   Solid state electronics

Probably the first observation on what we would now call a semi-conductor was made by Faraday in 1833, when he noticed that the electrical conductivity of solid silver sulphide increased when he raised its temperature. The 'cat's whisker', probably the first form of rectifier to be used as a detector of radio waves, also falls in this category, and from it developed the silicon diode which was so important in the development of radar in the early days of the Second World War. But the step which revolutionized electronics really occurred in 1947 at the Bell Telephone Laboratories in the USA, when Bardeen, Brattain and Shockley found that if two point contacts were placed a few micrometres apart on a specimen of germanium, the current flowing between one point and the germanium 'base' has a strong effect on the current flowing between the base and the other point. The device was shown to be capable of current, voltage and power amplification, and was named the *transistor*. From this discovery have developed many thousands of types of transistor, capable of replacing nearly all the functions of the thermionic valve, other devices which have no 'thermionic' counterpart (e.g. the triac), and, most important, *integrated circuits* – complete circuits made at a cost similar to that of making a single diode or transistor, and having almost any desired function.

We will look in this chapter at the methods of operation of the junction diode and a simple kind of transistor, and show how the transistor is used as an amplifying element. Then we will discuss integrated circuits and we will find that it is not necessary to understand the details of their operation but that by regarding them just as 'black boxes', or specialized building bricks, we can construct sophisticated circuits with amplifying or 'logic' functions.

p- and n-type semiconductors

A perfect crystal of, say, germanium has a simple regular structure (Fig. 12.1). Each atom makes four electronic bonds with neighbours. But if, say, one part in 10^6 of an element such as phosphorus (charge per ion 5) is present in the germanium melt when it starts to crystallize, the

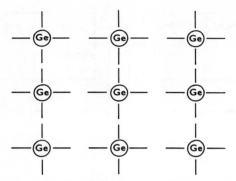

Fig. 12.1. A perfect crystal lattice of germanium atoms.

final crystal will have an extra electron near the site of the phosphorus atom. Such a material is called an *n*-type semiconductor, because conduction through it depends mainly on such 'spare' electrons.

Figure 12.2 shows a similar situation, the impurity this time being boron, which has only three electronic bonds. Thus there is a site in the crystal where an electron is missing; such a site is called a 'hole', and can move around in the crystal in the same way as an electron can. This kind of semiconductor is called '*p*-type'.

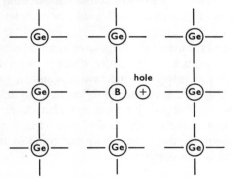

Fig. 12.2. Some *p*-type germanium (due to the addition of boron, an acceptor impurity).

The p–n junction

A junction between *p*- and *n*-type semiconductors has the very useful property of rectification, i.e., it presents a very high resistance to current flow in one direction, but a very low resistance to current flow in the opposite direction. This is illustrated in Fig. 12.3 (*a*) and (*b*).

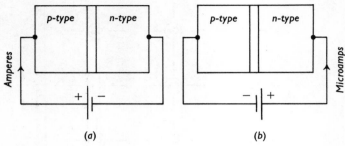

Fig. 12.3

When the junction is forward-biased, as in (a), holes from the p-type material easily acquire the energy necessary to cross into the n-type; as soon as they have done so, their 'lifetime' – the time before they are reoccupied by electrons – obviously decreases greatly; but the crossing of charge has occurred. Likewise, electrons from the n-type, under the repulsion from the external source of p.d., cross into the p-type material; they, too, soon recombine.

When the material is reverse-biased, as in (b), the external p.d. is trying to keep the 'holes' in the p-type, and the electrons in the n-type, so 'diffusion' of these respective carriers across the boundary is made greatly less, probably even than in the 'unbiased', or completely unconnected, junction.

p–n junctions intended for use as rectifiers have many advantages over the thermionic diode. They are much smaller, cheaper to produce, need no heater current, are not subject to mechanical damage, and handle currents measured in amperes, as compared to tenths of an ampere for a typical thermionic diode. The p–n junction can also be produced in specialized forms to handle the rectification of radio-frequency signals (see detection), or to switch radio signals from one route to another, or perform logic operations in a computer.

The transistor

A transistor of the junction type is shown in Fig. 12.4. It is a sandwich of n-type material between two slices of p-type, though many transistors are the other way round with p-type material enclosed by n-type. The emitter–base junction is shown forward biased, i.e. holes can pass easily from emitter to base. The base–collector junction is shown reverse biased – holes in the collector will not pass into the base against the potential provided by E_2. (When, for simplicity, we concentrate on the

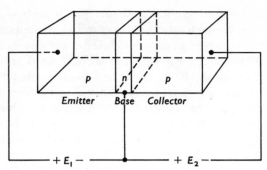

Fig. 12.4. The basic *p–n–p* transistor.

charge carried by just one type of carrier, holes or electrons, you should reason for yourself how the carrier which is *not* mentioned will behave. Often the two types of carrier contribute symmetrically with holes going one way and electrons the other.)

Unlike the thermionic valve, whose operation can be well understood by classical physics, the transistor requires the full machinery of quantum mechanics for a satisfactory explanation. However, the following discussion should help you to grasp its function.

The holes flowing out of the emitter, in Fig. 12.5, fall into two main

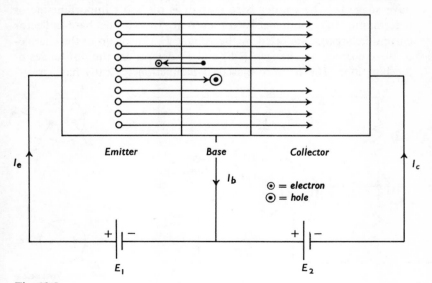

Fig. 12.5

categories: those which recombine with electrons in the base region and those which escape through into the collector region. Suppose on average a hole, once in n-type material, diffuses a distance L before disappearing in a recombination with an electron. In a transistor the width of the base region, W, is made very small so that W is much smaller than L. If on average all the holes are absorbed in a distance L, then a fraction W/L will be absorbed in the distance W. Thus the ratio of charge stopped in the base, to the charge carried by the holes into the collector is W/L. So the hole contribution to the emitter–base current is W/L times their contribution to the emitter–collector current.

We must now consider the other contributions to the emitter–base, base–collector and emitter–collector currents. Fortunately they are all small. The contributions due to electrons crossing from base to collector or holes from collector to base are negligible because, as we have said, the base–collector junction is reverse biased. The flow of electrons from the base into the emitter turns out, on full theoretical analysis, to be of the same order as the hole contribution to the base current. Thus the ratio of the collector current to the full base current is proportional to L/W which is typically of order 100. A small change in the base current will produce a much larger change in the collector current and so current amplification is achieved.

The emitter–base current is controlled by the voltage E_1, and as we have suggested the emitter base current is the most important factor determining the emitter–collector current. Indeed the base–collector current is largely unaffected by the voltage E_2. The ratio of the changes of these currents to changes in E_1 and E_2 determines the resistances of the junctions. The forward-biased E–B junction typically has a resis-

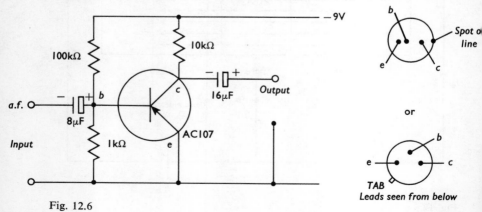

Fig. 12.6

tance of 1000 Ω, while the reverse biased B–C junction has a resistance of 100 000 Ω.

By following the collector with a high resistance, Fig. 12.6, voltage amplification is achieved. A small voltage change at the base causes a small change in the base current and hence a large change in the emitter–collector current, and a large voltage change across the 10 kΩ output resistor. The circuit is suitable for use as an audio–frequency voltage amplifier. It is in what is called the 'common emitter' mode, i.e. both input and output have the emitter as one connection. Compare it to the triode amplifier circuit. An important difference is that the input impedance is much smaller (about 10^3 instead of 10^6 Ω), and the voltage swing which can be handled is, for typical a.f. transistors, much smaller– say 3 V instead of 100 V.

Provided the low input impedance is kept in mind, many transistor circuits can readily be compared to the valve equivalent. The most obvious advantages are smallness, absence of the need for a heater supply, immunity to mechanical shock, and low voltage operation.

The 100 kΩ/1 kΩ potential divider biases the emitter-base junction in the forward direction, with a very small voltage.

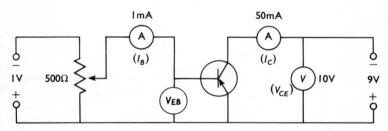

Fig. 12.7

As the transistor is most commonly used in 'common emitter' connection, a simple experiment to find its characteristics when connected this way is instructive. Figure 12.7 shows a suitable circuit. It helps when taking readings if the four meters are set out in easily visible order, and labelled as in the diagram.

The 'input characteristic' is a graph of I_B and V_{EB}, V_{CE} being kept constant. Do not allow the I_C meter to go off the scale.

The 'output characteristic' is a graph of I_C against V_{CE} for various fixed values of I_B. After setting a new value of V_{CE}, it will usually be necessary to reset I_B to the chosen value.

The usefulness of the transistor as a 'current amplifier' is often called its β or h_{fe}. Expressed mathematically it is

$$\left(\frac{\partial I_C}{\partial I_B}\right)_{V_{CE}}$$

that is, the change in collector current I_C corresponding to a certain change in base current I_B, the collector–emitter voltage being kept constant. It can be found directly using this circuit, or from the input and output characteristics. Typical values are 20 to 200.

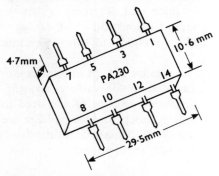

Fig. 12.8. An integrated low-level amplifier.

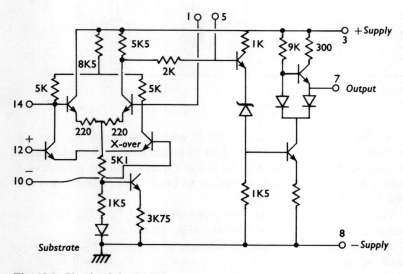

Fig. 12.9. Circuit of the PA230.

Integrated circuits

Photographic methods are now commonly used to produce thousands of identical transistors simultaneously. Such methods can equally well be used to produce 'microcircuits' containing many transistors, diodes, resistors and capacitors. At present, suitable inductances cannot readily be incorporated in integrated circuits, but the need for them can often be bypassed by altered circuit design.

A typical a.f. amplifier in integrated circuit form is shown in Fig. 12.8, and its circuit in Fig. 12.9. The external components and connections needed to use it as a low-power, high-gain, a.f. amplifier are shown in Fig. 12.10.

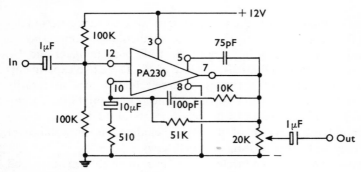

Fig. 12.10. An a.f. amplifier, voltage gain: 100.

'Systems' electronics

With integrated circuits such as the above available for less than £1, many users and designers of electronic devices will be more concerned with the assembling of 'systems' from 'black boxes', rather than in the design of the 'boxes' themselves. In this section we will see a few ways in which a small number of 'black boxes', having known properties, can be put together to make a 'system'.

The first 'black box' we shall look at is called a 'basic unit' (BU) or 'multifunction module'. Like the other modules, and like many commercial integrated circuits, it requires a steady 5 V d.c. supply. It has 3 'inputs', 2 marked as resistive, and the third as capacitive. There are 2 outputs, 'direct' and 'capacitive'. The emphasis is on knowing what can be done with the unit, *without* knowing what is in it.

First, a graph can be made of input voltage against output voltage, using the circuit of Fig. 12.11.

Using the graph, the student should first reason out what would be the

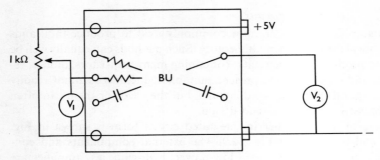

Fig. 12.11

waveform of the output if the input voltage were to vary sinusoidally between 0 and $+5$ V, and then check his prediction by connecting an audio-frequency oscillator in place of the 1 kΩ potentiometer and V_1, and an oscilloscope (Y terminals) in place of V_2. These two experiments show the capabilities of a transistor circuit (details not necessarily known) as (a) a switch, (b) a 'squarer', and (c) an amplifier. To emphasize (c), reduce the input from the a.f. to about 0·1 V peak.

The amplifying function of the BU, including the importance of biasing, may be further investigated as follows.

The capacitive input is connected to the 'live' of an a.f. generator, and one resistive input to the slider of a 1 kΩ potentiometer connected across the supply. An oscilloscope shows the output at the capacitive terminal, and the importance of correct 'biasing' is seen when the potentiometer setting is varied. Too large an input or too high a bias voltage leads to the squaring of the waveform already mentioned.

The next 'black box' to be studied is the Lamp Indicator Unit. It has one input; if this is at $+5$ V the lamp lights and if it is at 0 V the lamp is out. These conditions are often referred to as 'high' or 'low' respectively. By connecting the input to a potentiometer and voltmeter (same input

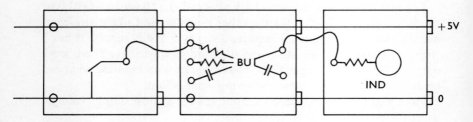

Fig. 12.12

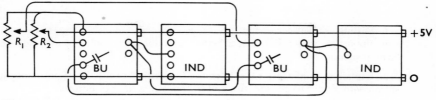

Fig. 12.13

as Fig. 12.11), the voltage at which the lamp first lights, and the voltage at which it appears fully lit, can be found.

The Multifunction Module (BU) can now be used to demonstrate the 'NOT' gate found in 'logic' circuits. The circuit is shown in Fig. 12.12. The lamp lights if there is NOT any input.

A slow-running multivibrator can be made by connecting 2 BU's and 2 indicators as in Fig. 12.13. The extra resistances R_1 and R_2 adjust the time for which each lamp is 'on' in the cycle.

Mention should be made of two other 'black boxes'. The 'bistable' has an output either 'on' or 'off'; it can be set either to change state whenever a pulse is applied at the input, or to change back to a 'preferred' state after a preselected time. The 'AND' gate module will provide an output only if both inputs receive a signal at once.

Once you know the functions of these few 'electronic building bricks', you can devise many interesting systems using only a few of them. For example, Fig. 12.14 operates the three indicators in the sequence of red, yellow and green traffic lights.

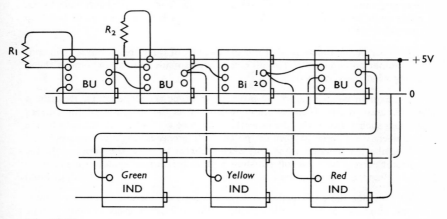

Fig. 12.14

Questions

1. Distinguish between conductors, semiconductors and insulators. How would you support the contention that a better name for the middle group would be 'semi-insulator'?

2. Explain in simple terms the difference in electrical properties of pure silicon, silicon with a trace of phosphorus, and silicon with a trace of boron.
A 'hole' has been compared to a vacant seat in an otherwise full cinema. Develop the analogy.

3. Explain the terms 'forward-biased' and 'reverse-biased' as applied to a single p–n junction (a diode). Why does such a device rectify? What are its advantages and disadvantages, compared to (thermionic) valve rectifiers?

4. Describe the behaviour of 'holes' in a p–n–p junction transistor. Devise a circuit for finding how its base–collector current depends on its base–emitter current when the junctions are respectively reverse and forward biased. Give typical component values and likely results.

5. Compare and contrast a triode valve (voltage) amplifier and a transistor (common emitter) amplifier.

6. A certain transistor has a β of 50. By how much does its collector current change when the base current is changed by 3 μA?
If this current passes through a collector resistance of 180 Ω, by how much does the voltage across the resistor change? (Neglect the change in collector voltage, but state what qualitative effect it would have if taken into account.)

7. What do the terms 'input characteristic' and 'output characteristic' mean, applied to a transistor in 'common emitter' connection? Sketch graphs of typical values for a small, low-power, a.f. amplifying transistor.

8. Discuss the relative importance of (a) being able to design a circuit using separate transistors to perform a specified single function, (b) being able to use ready-made circuits, without having any details of how they work, to build up complex systems.

9. Describe the functions, but not the circuits, of three 'black boxes' you consider important, and sketch two different systems made from all or some of them, saying what the systems do.

10. Explain what is meant by p-type and n-type semiconductors. Describe a p–n junction diode. Draw a graph which shows the variation of the current through such a diode with the potential difference across it, and explain why the diode behaves differently when the potential difference across it is reversed.
Describe the junction transistor. Sketch curves to show the variation of the collector current with the collector-base voltage for various values of the emitter-base voltage, and explain their form.   (O & C 1968)

11. Describe the construction and the principle of operation of *either* a thermionic diode *or* a p–n junction diode. Draw the appropriate current voltage characteristic and explain the physical processes which account for its form.

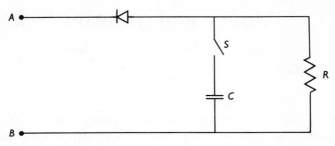

Fig. 12.15

The diode in Fig. 12.15 offers no resistance to the passage of current flowing in one direction but allows no current to pass in the opposite direction.

A voltage which varies sinusoidally with time is applied across the terminals A and B from a generator of negligible internal impedance. Draw the voltage waveforms which appear across the resistor (a) when the switch S is open, (b) when the switch S is closed and (c) when the switch S is open and the ideal diode is replaced by a real diode of the type whose current voltage characteristic you have drawn. In (b) assume that $RC \gg 1/f$ where f is the frequency of the voltage source.

(O & C 1971)

13   Atomic and nuclear physics

In the last decade of the nineteenth century three fundamental discoveries, the electron, X-rays and radioactivity, provided entry into a
new and unsuspected world, that of the atom. The growth of numerous
centres of scientific training and research, and later the stimulus of war,
enabled these discoveries to be developed and exploited with unexampled
swiftness.

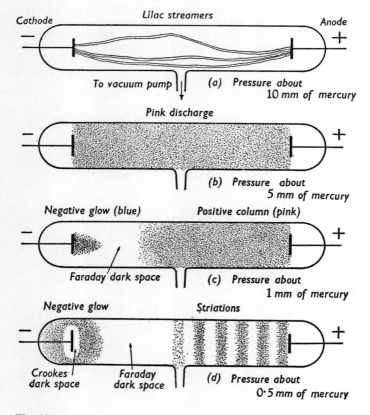

Fig. 13.1

The discharge of electricity through gases

The electron was discovered from the study of the discharge of electricity through rarefied gases. The preliminary stages of the discharge, as the pressure of a gas is reduced, are rather striking and beautiful. Figure 13.1 represents a long glass tube having electrodes at its ends. The electrodes are connected to the terminals of an induction coil which acts as a unidirectional source of several thousand volts. As the pressure of the air is reduced by a pump, lilac streamers first pass between the electrodes. Then the tube becomes filled with a pink glow. The colour varies with the gas in the tube: neon gives a red glow, argon pale blue, mercury green and helium yellow-white. As the pressure of the air is further reduced, the pink glow, called the positive column, shrinks towards the anode, and a dark space, called the Faraday dark space, together with a blue negative glow round the cathode appear. At a still lower pressure the positive column breaks up into striations, and shrinks still further, while the Crookes dark space appears between the cathode and the negative glow. Eventually at a pressure of about 0·02 mm the positive column, the Faraday dark space and negative glow have disappeared, and the Crookes dark space has expanded to fill the whole tube. The glass of the tube then fluoresces a green colour.

These complicated phenomena will be considered again later.

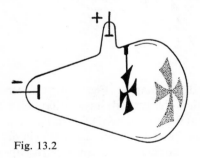

Fig. 13.2

Cathode rays

The electron was discovered by the investigation of the Crookes dark space and the following experiments refer to gas pressures such that the Crookes dark space fills the tube. If an obstacle such as a mica cross is placed in front of the cathode, a shadow of the cross is thrown on the end of the tube (see Fig. 13.2). Certain minerals, when placed in front of the cathode, fluoresce with brilliant colours. Something appears to be emitted from the cathode and this invisible radiation was termed

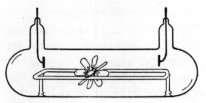

Fig. 13.3

cathode rays. It was found that cathode rays would pass through an aluminium window. German investigators believed that they consisted of penetrating invisible light; English investigators, on the other hand, believed that they consisted of tiny particles of negative electricity. The latter view was substantiated by the three following experiments. A light mica paddle wheel is made to rotate on rails by the impact of the cathode rays (Fig. 13.3); when the polarity of the electrodes is reversed, the direction of rotation of the paddle wheel is changed. Again, the cathode rays, made visible by a fluorescent, zinc sulphide screen fitted

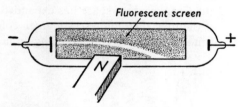

Fluorescent screen

Fig. 13.4

inside the tube, are deflected by a magnet (Fig. 13.4); by Fleming's left-hand rule it can be seen that the cathode rays behave like a flow of negative charge. Finally, and this experiment of Jean Perrin in 1895 was crucial, the cathode rays can be collected on a metal cylinder, called a Faraday cylinder, which is connected to an electroscope, thereby proving directly that the cathode rays carry a negative charge (Fig. 13.5).

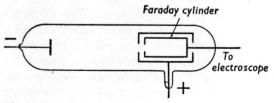

Faraday cylinder

To electroscope

Fig. 13.5

The anode is perforated and surrounds the Faraday cylinder to shield the latter from electrostatic disturbance.

Sir William Crookes, as early as 1874, suggested that the cathode rays represent 'a fourth state of matter' and remarked with extraordinary insight, 'I believe that the greatest scientific problems will find their solution in this borderland, and even beyond it; it seems to me that here we have reached the ultimate realities.'

Measurements of e/m and v for the cathode particles

Cathode rays were studied in various parts of the world, notably in the Cavendish Laboratory, Cambridge, under the direction of J. J. Thomson. In 1897 Thomson measured the ratio of the charge to the mass of the particles, and also their velocity, by deflecting them in magnetic and electric fields. His apparatus is represented in Fig. 13.6. A

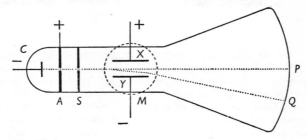

Fig. 13.6

fine beam of cathode rays is produced by means of a cathode C, a perforated anode A and a slit S, and makes a bright spot P on the end of the tube. X and Y are metal plates between which an electric field can be applied and the dotted circle M represents two coils, one on each side of the tube, which enable a magnetic field to be applied at right angles to the paper.

Suppose a magnetic field only is applied. The cathode rays are deflected and the bright spot moves from P to Q. (If the magnetic field is reversed then the deflection will also be reversed.) The force exerted on a charged particle while passing through the uniform magnetic field is always at right angles to the magnetic field and to the direction of motion of the particle. Thus the particle will move in an arc of a circle while passing through the magnetic field. If e is the charge and v the velocity of the particle, the force exerted on it by the magnetic flux density, B, is Bev (see p. 53). This force will equal the centrifugal force

SIR J. J. THOMSON   LORD RUTHERFORD

SIR J. J. THOMSON (1856–1940) *was the discoverer of the electron and, in the words of Sir William Bragg, 'more than any other man was responsible for the fundamental change in outlook which distinguishes the physics of this century from that of the last'. At the turn of the century J. J., as he was affectionately called, attracted to the Cavendish Laboratory a brilliant team of research workers whom he directed in investigations connected with the discharge of electricity through gases. He would visit the room of each worker daily, leaving open the door, and his loud voice with its trace of a Lancashire accent would resound throughout the building; despite his own lack of manual skill he often displayed an uncanny insight into the working of their apparatus. In 1918 he became Master of Trinity College, Cambridge, and was succeeded as Cavendish Professor of Experimental Physics by Rutherford, a former pupil. He was buried in Westminster Abbey near to the tombs of Newton, Darwin, Kelvin and Rutherford.*

LORD RUTHERFORD (1871–1937), *a New Zealander, was perhaps the greatest physicist since Newton. He was foremost in investigating radioactivity, and in 1903 put forward, with Soddy, the disintegration theory to account for it. In 1911 he propounded the nuclear theory of the atom, based on his experiments on the scattering of α-particles, and in 1919 he effected the first artificial disintegration of an atom. Sir James Chadwick has said of him: 'It is amazing that one man could so transform physics by his own effort, a man with no great mathematical equipment, good but not remarkable even in experimental technique. Nor had he an acute or subtle mind; no, his mind was like the bow of a battleship – there was so much power and weight behind it, it had no need to be as sharp as a razor. He brushed aside all irrelevancy and went straight to his mark.'*

mv^2/r, where m is the mass of the particle and r the radius of the circle in which it moves.

$$\therefore \ Bev = \frac{mv^2}{r},$$

$$\frac{e}{m} = \frac{v}{rB}.$$

The value of r is known from the distance PQ and the dimensions of the apparatus.

Now suppose that an electric field is applied between X and Y in such a direction and of such strength that it brings the spot of light back from Q to P, thus annulling the deflection due to the magnetic field. The force on each charged particle, due to the electric field strength E, is Ee and this must equal the force due to the magnetic field.

$$Ee = Bev,$$

$$v = \frac{E}{B}.$$

Knowing E and B, v can be calculated. Thomson found that its value was of the order of $\frac{1}{10}$th the velocity of light, and that it varied with the p.d. between the anode and the cathode.

Knowing v it is possible to obtain e/m from the above equation. Thomson found that the value of e/m is constant for all gases and electrodes.

Assuming e to be the same as the smallest charge carried by an ion in electrolysis (see p. 149) he calculated that the particles forming the cathode rays, now called electrons, had a mass only $\frac{1}{1840}$ of that of the lightest known atom, hydrogen.

This was an astonishing result at the time, and an alternative interpretation, preferred by some physicists, was that e was 1840 times larger, rather than that m was 1840 times smaller, than the known values for the hydrogen ion. Accordingly Thomson and Townsend devised a method of finding e separately.

Determination of the charge of the electron

Townsend's method was to determine the number of drops in a charged fog formed on negative gaseous ions, and to measure the total charge. The charge on each drop could then be calculated. Gaseous ions are produced by molecules losing or gaining electrons and hence carry the electronic charge.

Water drops do not form in slightly supersaturated air unless it contains nuclei like specks of dust on which the vapour can condense. Gaseous ions act as nuclei for condensation and negative ions are more efficient in this respect than positive. Thomson, in an improved version of Townsend's experiment, ionized a gas mixed with saturated water vapour by exposing it to X-rays and caused a fog to form on the negative ions by sudden expansion. The fog was allowed to settle to the bottom of the apparatus where its total charge was measured with a quadrant electrometer, and also the total volume of the drops was found. The size of each drop could be calculated from the rate of fall, measured by a small telescope focused on the top edge of the fog as it settled. Thomson obtained an approximate value for *e* which completely confirmed his view that the electron had a mass very much less·than the lightest atom.

The American physicist, Millikan, modified the fog method and in 1906 made a very accurate determination of *e*. There were several serious sources of error in the earlier experiments. The fog droplets were not all of the same size and measurements were made only on the smallest, which fell most slowly. Moreover, the individual droplets did not remain the same size; the smallest tended to evaporate and the largest to grow.

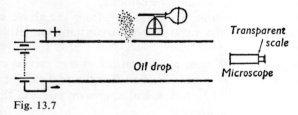

Fig. 13.7

Millikan formed drops of non-evaporating oil instead of water and observed the motion of a single drop instead of a whole cloud. His apparatus is indicated in Fig. 13.7. Oil drops formed by a spray are allowed to fall between two plates. They are brightly illuminated and observed through a low-power microscope. The plates are first earthed and the rate of fall of a drop timed by a stop-watch. From this the mass of the drop can be found. Stokes's formula, $F = 6\pi\eta av$, gives the force of resistance F to the fall of the drop, where η is the viscosity of the air, a is the radius of the drop and v is the velocity. When the drop has reached its terminal velocity, F is equal to the weight of the drop minus the upthrust of the air.

$$\therefore \ 6\pi\eta av = \tfrac{4}{3} \pi a^3(\rho_1 - \rho_2)\, g,$$

where ρ_1 and ρ_2 are the densities of the oil and the air respectively. Hence the radius of the drop and its mass can be found.

Drops formed by a spray are usually found to be charged but, if necessary, they may be given a charge by directing on to them a beam of X-rays. The fall of a negatively charged drop can be arrested by giving the upper plate a positive potential. The drop can be made to rise and fall at will by adjusting the p.d. between the plates and a single drop can be kept under observation for hours.

Suppose the drop is held stationary by a p.d. of V, and the plates are d apart. The electric field strength between the plates is V/d. Thus the force on the drop of charge e is Ve/d. This force must equal the apparent weight of the drop, which is known from the rate of fall between uncharged plates. Hence the value of e can be calculated.

Millikan obtained the value $e = -1.60 \times 10^{-19}$ C for the smallest charge carried by a drop, although drops sometimes acquired charges exactly two or three times this value.

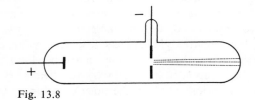

Fig. 13.8

Positive rays

In 1886 Goldstein discovered that if a hole is bored in the cathode of a discharge tube a reddish streamer of radiation can be seen on the further side of the cathode (Fig. 13.8). He called this radiation 'canal rays' because it passed through a canal in the cathode. In 1898 Wien deflected the radiation with a powerful magnet and deduced that it consisted of positively charged particles; hence the usual name, positive rays.

In 1910–11 J. J. Thomson measured the deflection of positive rays when subjected simultaneously to a magnetic and an electric field. From the values obtained for e/m, assuming e to be equal in magnitude but opposite in sign to the charge of the electron, he found that the masses of the particles in the positive rays are those of atoms or molecules and that they vary with the nature of the gas used. The particles were therefore taken to be atoms or molecules of the gas which had lost one or more electrons, i.e. positive ions.

Explanation of the phenomena of the discharge tube

The conductivity of a gas, like that of a liquid, is due to positive and negative ions travelling in opposite directions. A discharge can start only if the field is sufficiently intense to ionize the gas, unless a separate source of ionization, such as X-rays, is used. The phenomena of the discharge tube can be accounted for by the production of ions by collision in various parts of the tube, luminosity being a result of ionization. The clue needed to solve the problem was the very different fall of potential in various parts of the tube, determined by inserting subsidiary electrodes.

Consider the state of the discharge in the bottom of Fig. 13.1. The fall of potential in the Crookes dark space is very much greater than in any other part of the tube. Here the positive ions acquire their maximum speed and, as already described, if allowed to pass through a hole in the cathode, they will ionize the gas on the other side, causing a reddish stream of illumination.

The negative ions or cathode rays, caused probably by the bombardment of the positive ions on the cathode, move with increasing speed in the opposite direction and collide with molecules of the gas, causing the ionization in the negative glow. They lose energy by collision, becoming incapable of further ionization, and the Faraday dark space results. On being speeded up again by the field they acquire sufficient energy to cause ionization in the positive column. The striations are accounted for by alternate ionization and subsequent increase of speed of the negative ions to produce fresh ionization.

Isotopes

An interesting discovery was made by Thomson during his investigation of positive rays which recalled a discarded hypothesis put forward by Prout in 1815, that the atomic weights of all the elements should be exact integers. The gas neon, of atomic weight 20·2, was found to consist of two kinds of atoms, of masses 20 and 22. The existence of atoms, having identical chemical properties but different masses, was earlier postulated by Soddy from his experiments on radioactivity, and he called them *isotopes*, the Greek derivation of the word indicating that they have the same place in the periodic table of the elements.

Aston refined Thomson's method and undertook the task of examining all the elements. When his apparatus was perfected in 1919, scarcely a week passed without the announcement of new isotopes. There are more than 250 isotopes of the 92 natural elements. The most complex element is tin with no fewer than ten isotopes.

One of the most important isotopes is heavy hydrogen of mass 2 discovered by Urey in America in 1932 and now called deuterium. Its compound, heavy water or deuterium oxide, has been isolated by repeated electrolysis of water; the light hydrogen comes off first leaving behind a progressively richer mixture of deuterium oxide which, when pure, has density of 1100 kg m^{-3} and freezing and boiling points of 277 and 374·6 K respectively.

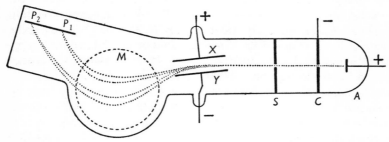

Fig. 13.9

Aston's mass spectrograph

Aston's apparatus, called a mass spectrograph, is shown in Fig. 13.9, A is the anode and C the cathode. The positive rays emerging from the fine hole in the cathode pass through a narrow slit S; they are deflected downwards by an electric field between the plates XY and upwards by the magnetic field between two poles M of a powerful electromagnet. Aston's chief improvement on Thomson's method was the focusing of the rays; all particles of the same mass are caused to converge to a focus. The slower particles are deflected more than the faster ones by the electric field since they are under its influence for a longer time. The magnetic field gives a similar effect but in the opposite direction, so that, by adjustment of the field strength, the dispersion can be annulled. Particles having different masses are focused at different points and thus images P_1, P_2 etc., are produced on a suitably placed photographic plate for each of the isotopes which are present.

The type of photograph obtained is shown in Fig. 13.10. From the

Fig. 13.10. Mass spectrogram of zinc's five isotopes, 64, 66 and 68 are in greatest quantity, 67 in less and 70 in very small quantity.

dimensions of the apparatus and the strengths of the fields, the masses of the particles can be calculated.

To obtain positive ions of elements which are not normally gaseous Aston used an anode having a cavity packed with a paste containing one of the compounds of the element.

Aston's original apparatus had an accuracy of 1 in 1 000 which was later improved to 1 in 10 000. The masses of all isotopes are not exact whole numbers and the slight discrepancies have proved to be of great theoretical importance. Einstein obtained the following simple relation between a mass m and its equivalent energy E: $E = mc^2$, where c is the velocity of light. Thus if 1 kg of matter could be annihilated the energy released would be $1 \times (3 \times 10^8)^2 = 9 \times 10^{16}$ joules, which is the energy obtained from burning about 4 000 000 tons of coal!

The helium nucleus, of mass 4·002 80, is now believed to be made up of two protons, each of charge e and mass 1·008 1, and two neutrons, each of zero charge and mass 1·008 9. Thus the separate components of the helium nucleus have a mass $2 \times 1·008\ 1 + 2 \times 1·008\ 9 = 4·034\ 0$, which is considerably more than the mass of the helium nucleus. Astronomers now account for the radiation of the sun and stars by the annihilation of mass during the building up, from protons and neutrons, of helium and more complex elements. It is calculated that the sun is losing 4 000 000 tons of mass per second, but would take 15×10^{12} years to radiate its whole mass. The atomic bomb and the nuclear reactor also derive their energy from the annihilation of mass.

X-rays

In 1895 Röntgen, Professor of Physics at Würtzburg, while experimenting with a discharge tube, discovered that photographic plates wrapped in black paper and left near the tube, were fogged. His laboratory assistant noted that a silhouette of a key resting on the box of plates appeared on the plates after development. The tube was found to be emitting radiation of a penetrating type which Röntgen called X-rays. Sir William Crookes, who was also bothered by the fogging of his plates when photographing discharges at low pressures, moved them to another room and so, while saving his plates, possibly missed discovering X-rays.

X-rays are produced whenever fast moving electrons are stopped suddently. The earliest type of X-ray tube, known as a gas tube, is shown in Fig. 13.11. The gas in the tube is at such a pressure that the Crookes dark space fills the tube. The cathode C is concave in order to focus the cathode rays on to a target or anti-cathode a.c. The anti-cathode may

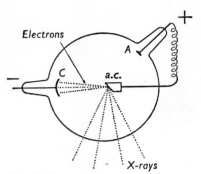

Fig. 13.11

also be used as the anode, but it is found that a separate anode A gives a steadier discharge. The pressure of the gas in the tube gradually decreases with use owing to occlusion of gas by the electrodes; this causes the X-rays produced to become less intense but 'harder' or more penetrating. 'Softening' devices for releasing a little more gas into the tube when necessary are usually provided.

A greatly improved type of tube was invented by Coolidge in 1913 (Fig. 13.12). The tube is exhausted as highly as possible so that no discharge would pass through it if it were used as a gas tube. The electrons are emitted by a tungsten filament C and the rate of their emission, which determines the intensity of the X-rays, can be controlled by the heating current through the filament.

The hardness of the X-rays is controlled separately by the p.d. between A and C. X-rays are electromagnetic waves like light but of much shorter wavelength. Their hardness depends on their frequency and this is greater the faster the speed of the electrons hitting the target. The phenomenon is a kind of reverse photoelectric effect and a similar relation holds:

$$Ve = hf,$$

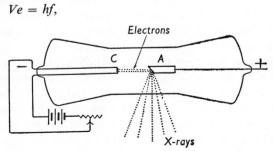

Fig. 13.12

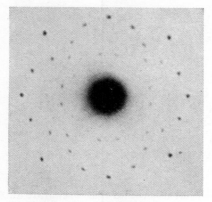

Fig. 13.13. Laue photograph of rock salt.

where V = p.d. through which the electrons fall, e = charge of the electron, h = the Planck constant, f = maximum frequency of the X-rays emitted.

Only about 1% of the energy of the electrons is converted into X-rays. The rest of the energy appears as heat, and hence the target is made of a metal of high melting-point, such as tungsten or molybdenum, embedded in a solid block of copper which is a good thermal conductor.

Until 1912 the true nature of X-rays was unknown. Some believed that they were waves like light, and others, among whom was Sir William Bragg, who was later to measure their wavelength, believed that they were particles. Their known properties were as follows:

(1) They travel in straight lines and are undeviated by magnetic and electric fields.

(2) They affect a photographic plate and cause luminescence on a screen of zinc sulphide or barium platinocyanide.

(3) They are very penetrating. Their absorption depends on the density and thickness of the matter through which they pass. Thus bones absorb them more readily than flesh; a thickness of 1 mm of lead is opaque to X-rays of moderate hardness.

(4) They liberate electrons from air (ionizing it) and from certain metals. Thus they cause a charged electroscope to become discharged.

Early attempts to diffract X-rays with the most closely ruled gratings were unsuccessful, and this was the fact which weighed most heavily with the supporters of the particle theory. In 1912, however, von Laue, of the University of Munich, conceived the idea that crystals, with their regularly spaced atoms, might behave as gratings and produce diffrac-

tion patterns with X-rays. The experiment was tried and was immediately successful (Fig. 13.13).

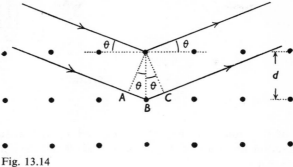

Fig. 13.14

Determination of the wavelength of X-rays

In 1913 Sir William Bragg and his son, Sir Lawrence Bragg, devised a
comparatively simple method of measuring the wavelengths of X-rays
by reflecting them from the surface of a crystal. In the method of von
Laue the rays are transmitted through the crystal.

Huyghens explained the reflection of light waves by assuming that
each point on the reflector struck by the incident waves acts as a source
of secondary waves and he showed that these secondary waves reinforce
each other when the angle of incidence is equal to the angle of reflection.
The atoms (or more strictly the ions) of a crystal act as sources of
secondary waves when struck by X-rays. Fig. 13.14 shows reflection
of X-rays at two successive layers of ions in a crystal. The lines with
arrows represent the direction of travel of the waves and hence are
rays. The complete reflected wave will be strong only if the reflections
from successive layers, probably from many millions of layers, reinforce
each other. This will occur if the difference in path, $AB + BC$, is an
integral number of wavelengths, i.e. if

$$2d \sin \theta = n\lambda,$$

where d = distance apart of the adjacent layers of ions, n = an integer,
λ = wavelength of X rays.

The Bragg X-ray spectrometer is shown in Fig. 13.15. The intensity of
the 'reflected' waves is measured by means of an ionization chamber,
consisting of a lead cylinder containing air in which is fitted an insulated
electrode. When a beam of X-rays enters the chamber it ionizes the air,
enabling a current to flow between the walls of the chamber and the
electrode, and causes a deflection of the gold-leaf electroscope. The

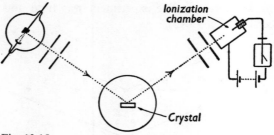

Fig. 13.15

crystal and the ionization chamber are rotated together at suitable rates and measurements made of the glancing angle θ and the corresponding ionization current.

It can be seen from the equation $2d \sin \theta = n\lambda$ that for X-rays of a particular wavelength λ there will be a series of angles θ corresponding to $n = 1, 2, 3$, etc., at which a strong beam is reflected. Thus spectra of several orders must be distinguished.

The value of d for rock salt (NaCl) can be calculated quite simply since the crystal form of the latter is known to be that a cubic lattice. Each cube in the lattice in Fig. 13.16 has an ion at each corner, eight in all, and in a crystal containing a large number of cubes each ion, except those on the surface, will be shared by eight cubes (like the centre ion in Fig. 13.16). Thus there will be, on an average, one ion per cube. Now a gram-molecular weight of rock salt, i.e. $23 + 35 \cdot 5 = 58 \cdot 5$ g, contains $6 \cdot 06 \times 10^{23}$ atoms, or $2 \times 6 \cdot 06 \times 10^{23}$ ions. The density of rock salt is $2 \cdot 18$ g per cm^3 and hence 1 cm^3 contains $\dfrac{2 \times 6 \cdot 06 \times 10^{23} \times 2 \cdot 18}{58 \cdot 5}$ ions. But for each ion there is 1 cube of the lattice of volume d^3, where d is the distance apart of the ions.

$$\therefore \quad \frac{2 \times 6 \cdot 06 \times 10^{23} \times 2 \cdot 18}{58 \cdot 5} \times d^3 = 1,$$

$$d = 2 \cdot 8 \times 10^{-8} \text{ cm} = 280 \text{ pm}.$$

Typical values obtained by the Braggs for strong radiation from palladium, using a crystal of rock salt, were

n	θ
1	5° 59′
2	12° 3′
3	18° 14′

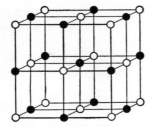

Fig. 13.16

Substituting for the first value,

$$\lambda = 2d \sin \theta$$
$$= 2 \times 2 \cdot 8 \times 10^{-8} \sin 5° \; 59'$$
$$= 0 \cdot 586 \times 10^{-8} \; cm = 58 \cdot 6 \; pm.$$

Once the wave-length of X-rays is known the method can be used to determine the spacing of the ions in crystals.

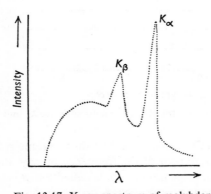

Fig. 13.17. X-ray spectrum of molybdenum.

Moseley's experiments

An X-ray tube emits a background of radiation, like the continuous spectrum of white light, on which is superimposed certain stronger radiations, like luminous line spectra (Fig. 13.17). The X-ray line spectra vary in wavelength with the element of which the anti-cathode is composed, and have been classified into a hard K radiation, and a softer L radiation. There is, in addition, a still softer M radiation in the case of the heavy elements.

In 1913 Moseley undertook the investigation of the X-ray spectra of the elements. He mounted a series of elements on a trolley in an X-ray tube and moved them so that they were struck in turn by a stream of electrons. The resulting X-rays were examined by a crystal. Moseley discovered that there is a regular shift of the spectral lines from element to element (Fig. 13.18). Each element can be given a number, beginning

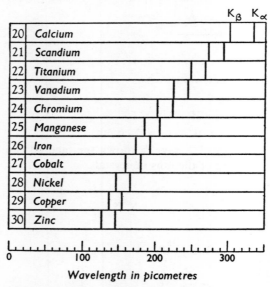

Fig. 13.18. X-ray spectra of the elements showing K_α and K_β lines.

with hydrogen 1, helium 2, lithium 3 and so on to uranium 92 which, if plotted against the square root of the frequency of, say, the K_α line, gives a straight line graph. This number, now called the *atomic number*, was found to be equal to the positive charge on the nucleus of the atom and also to the number of planetary electrons.

Moseley's discovery enabled the elements to be classified in a new and more satisfactory manner. If the elements are arranged in the order of their atomic weights certain anomalies occur, since elements with similar chemical properties do not fit into the periodic classification. Thus potassium precedes argon instead of following it and the same is true of nickel and cobalt. On arranging the elements in the order of their atomic numbers these anomalies are removed. Moseley found gaps in the table corresponding to numbers 21, 43, 61, 72 and 75. A search was made for the missing elements and some gaps were filled by scandium (21), hafnium (71) and rhenium (75).

Moseley's early death at the Gallipoli landing in 1915 was a tragic loss to science.

Radioactivity

X-rays cause certain substances, such as zinc sulphide or barium platinocyanide, to fluoresce. Henri Becquerel examined fluorescent substances to see whether they gave off radiation and in 1896 he found that uranium and its compounds, like X-rays, affect a photographic plate through a wrapping of paper. That the effect was not due to vapours was proved by inserting a sheet of glass between the uranium salt and the plate.

The phenomenon was investigated by Pierre and Marie Curie. The Austrian Government provided them with a ton of pitchblende, the ore of uranium, from Bohemia. In a leaky, abandoned shed near the School of Physics in Paris they began the laborious task of separating the radioactive components from the ore. First they isolated a new, active element which was associated with the bismuth in the ore. This was named polonium in honour of Madame Curie's native country. They realized, however, that there was a far more active element associated with the barium constituent and eventually they succeeded in extracting $\frac{1}{5}$ g of it from 1 tonne of ore. The new element, radium, had an activity about two million times that of uranium. Its supply of energy was prodigious and apparently inexhaustible; it provided enough heat every hour to raise its own weight of water from freezing to boiling point.

α, β and γ rays

The most acute and successful investigator of radioactivity was Rutherford, who left Cambridge in 1898 to take up a professorship at McGill University, Montreal. At this time practically all the radium in the world was in the possession of the Curies. Dr Giesel, however, put radium bromide on the market at £1 per milligramme and Rutherford bought 30 milligrammes; later the price rose to £12 per milligramme.

The radiation from radioactive substances is complex as can be shown by deflecting it in a magnetic field (Fig. 13.19). Part, called by Rutherford α-rays, is deflected in such a direction as to indicate that it carries a positive charge; another part, the β-rays which proved to be electrons, is deflected in the opposite direction; and a third part, the γ-rays which are X-rays of very short wave-length, is undeviated. The α-rays in Fig. 13.19 are less bent than the β-rays because they are heavier, and they are less dispersed because they are homogeneous as regards velocity, whereas β-rays are not.

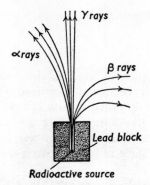

Fig. 13.19. Deflection of α- and β-rays by a magnetic field whose direction is into the plane of the paper.

In 1903 Rutherford measured the deflection of α-particles in a magnetic field and calculated the ratio of charge to mass. He deduced that the α-particles are charged atoms of helium, of mass 4 (taking the hydrogen atom as of mass 1) and of charge $+2e$ (where $-e$ is the charge of the electron). That the α-particles might be helium was suggested by the fact that radioactive minerals contain small quantities of helium. Rutherford obtained confirmation by collecting α-particles in an evacuated tube and showing that they gave the spectrum of helium when rendered luminous by an electrical discharge.

The velocity with which the α- and β-particles are emitted is enormous, that of the α-particles varying between $\frac{1}{100}$ and $\frac{1}{10}$ the velocity of light while that of the β-rays is sometimes as high as 99 per cent of the velocity of light. The γ-rays are given off whenever β-rays are ejected, which is understandable since X-rays are produced by a sudden change in the velocity of electrons.

The penetrating powers of the different rays in different media were closely studied: α-rays can penetrate only a small fraction of a millimetre of aluminium foil, β-rays penetrate several millimetres of aluminium, and γ-rays will penetrate several centimetres of lead. The range of α-particles in air at atmospheric pressure varies between 3 and 8 cm and can be used to identify different radioactive substances.

The Wilson cloud chamber

In 1912 C. T. R. Wilson devised an apparatus known as the cloud chamber for making visible the tracks of charged particles (Fig. 13.20). The space inside a vessel containing dustless air and saturated water

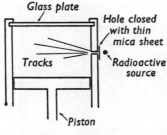

Fig. 13.20

vapour is made to expand suddenly by the rapid movement of a piston. Cooling and supersaturation result, and if the expansion is not too great (about one-third of the original volume), condensation will take place on charged ions without a general fog being formed.

An α-particle, as it ploughs its way through the air in a cloud chamber, knocks out electrons from thousands of molecules and leaves a trail of condensation on the ions produced (Fig. 13.21). A β-particle, being far lighter, is deflected by its collisions with the air molecules and produces

Fig. 13.21. Tracks of α-particles from thorium (C and C') showing the two groups of ranges 8·6 and 4·8 cm in air.

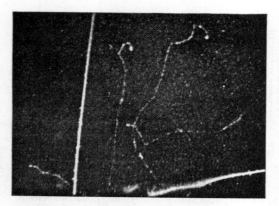

Fig. 13.22. Tracks of α- and β-particles in hydrogen. The thin curved tracks are those of β-particles.

far fewer ions than an α-particle (Fig. 13.22). X-rays cause electrons to be ejected from air molecules and produce tracks as in Fig. 13.23.

The cloud chamber was described by Rutherford as 'the most original and wonderful instrument in scientific history'. It is still of great value in atomic research, although it has been superseded, for use in high-energy physics, by the bubble chamber which uses nearly boiling hydrogen instead of supersaturated vapour. The pressure on the liquid is suddenly released so that it finds itself above its boiling point. Bubbles will then start to form but, as in the cloud chamber, they form preferentially around ions.

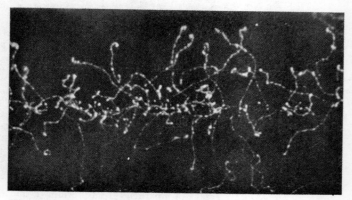

Fig. 13.23. Electron tracks produced by a beam of X-rays in a gas.

Theory of radioactive decay

In 1903 Rutherford and Soddy presented to the British Association their disintegration theory, that the atoms of a radioactive element explode with great violence, hurling out α- and β-particles and being changed in the process to a different element. The theory was so novel and startling that it aroused much opposition, including that of Lord Kelvin. Some other source of the immense energy of radioactive atoms was suggested, such as the absorption of hitherto undetected universal radiation.

The subsequent experimental unravelling of the intricate radioactive changes has, however, fully borne out the theory. There is some inherent instability in atoms of atomic number greater than 82, causing disintegration or decay. The process cannot be hastened or delayed by physical or chemical means. Its speed may be represented by the 'half-life period' which is the time required for half the element to break up and be transformed. The half-life period of radium is 1590 years; thus 1 g of radium is reduced to $\frac{1}{2}$ g in 1590 years, to $\frac{1}{4}$ g in a further 1590 years, and so on. The law is statistical. It is impossible to tell when an individual atom will decay, but in the case of a very large number of atoms it is known that a certain fraction of them will break up every second.

The table below shows the uranium–radium family, each element being produced by the decay of the element immediately above it in the table. The emission of an α-particle causes a loss of mass of 4 and a loss of charge of $+2e$. Since the atomic number of an element is the charge on the nucleus it is clear that the loss of an α-particle causes the element to be displaced two places down the atomic number table. When radium decays with the ejection of an α-particle it gives rise to the radioactive gas radon, a member of the family of inert gases. The disintegration may be represented as follows:

$$\require{mhchem} {}^{226}_{88}\text{Ra} = {}^{222}_{86}\text{Rn} + {}^{4}_{2}\text{He}.$$

The subscript at the bottom represents the charge or atomic number and the superscript at the top the atomic mass.

The emission of a β-particle during decay has a negligible effect on the mass of an atom, since the mass of an electron is only $\frac{1}{1840}$ of the mass of the hydrogen nucleus or proton, but it results in a loss of charge of $-1e$ which is equivalent to a gain of charge of $+1e$. Thus the element moves up one place in the atomic number table. For example,

$$ {}^{214}_{82}\text{Ra B} = {}^{214}_{83}\text{Ra C} + e.$$

Uranium–radium series of elements

Element	Abbreviation	Atomic number	Atomic mass	Particle ejected	Half-life period
Uranium I	U I	92	238	α	$4 \cdot 6 \times 10^9$ years
Uranium X_1	U X_1	90	234	β	245 days
Uranium X_2	U X_2	91	234	β	$1 \cdot 14$ min.
Uranium II	U II	92	234	α	3×10^5 years
Ionium	Io	90	230	α	$8 \cdot 5 \times 10^4$ years
Radium	Ra	88	226	α	1590 years
Radon	Rn	86	222	α	$3 \cdot 82$ days
Radium A	Ra A	84	218	α	$3 \cdot 05$ min.
Radium B	Ra B	82	214	β	$26 \cdot 8$ min.
Radium C	Ra C	83	214	β, α^*	$19 \cdot 7$ min.
Radium C′	Ra C′	84	214	α	10^{-6} s
Radium C″	Ra C″	81	210	β	$1 \cdot 32$ min.
Radium D	Ra D	82	210	β	$22 \cdot 3$ years
Radium E	Ra E	83	210	β	$5 \cdot 0$ days
Polonium	Po	84	210	α	139 days
Lead	Pb	82	206	—	Stable

* Ra C may pass to Ra D either by way of Ra C′ or of Ra C″. When Ra C ejects a β-particle Ra C is formed; when Ra C ejects an α-particle Ra C″ is formed.

There are two other families of radioactive elements, those of thorium and of actinium, making in all about 40 natural radioactive elements. The stable end-product of each family is an isotope of lead; ^{206}Pb for the uranium family, ^{208}Pb for the thorium family and ^{207}Pb for the actinium family.

The nuclear atom

The earliest hypothesis of the structure of an atom was the 'positive jelly model' of J. J. Thomson, in which electrons were regarded as embedded in a sphere of positive electricity.

In 1911 a 'planetary model', in which the electrons revolve like planets round a small, massive, positively charged nucleus resulted from Rutherford's experiments on the scattering of α-particles by gold or platinum foil. The apparatus is represented in Fig. 13.24; α-particles from a radioactive source, R, hit a thin metal foil, F, and are scattered. The number scattered in different directions can be counted by the scintillations which α-particles produce when they hit a fluorescent screen, S, at the focus of a low-powered microscope M.

Rutherford by this time had left Canada and was Professor of Physics in Manchester. He relates that one day Geiger came to see him and asked, 'Don't you think that young Marsden, whom I am training in radioactive methods, should begin a small research?' Rutherford replied, 'Why not let him see whether any α-particles can be scattered through a large angle?', feeling, at the same time, that there was little chance of this because of the immense speed and energy of the α-particles. Shortly afterwards Geiger reported that some α-particles had been observed which were scattered through angles greater than 90°.

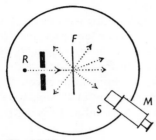

Fig. 13.24

'It was almost as incredible as if you had fired a fifteen inch shell at a sheet of tissue paper and it came back and hit you', said Rutherford. No diffuse positive jelly could account for such a result; the atom must contain something extremely hard and massive. By careful measurements of the scattering Rutherford was able to estimate that the diameter of the nucleus of an atom must be of the order of 10^{-13} cm. It was known that the diameter of an atom is about 10^{-8} cm. Thus the nuclear atom must consist largely of empty space.

The nucleus is now believed to consist of protons of mass 1 and charge $+1e$, and neutrons of mass 1 and charge 0. Surrounding the nucleus are planetary electrons of mass $\frac{1}{1840}$ and charge $-1e$. The constitutions of the three lightest elements are as follows:

	Nucleus	Planetary electrons
Hydrogen	Charge 1 mass 1 (1 proton)	1
Helium	Charge 2 mass 4 (2 protons and 2 neutrons)	2
Lithium	Charge 3 mass 7 (3 protons and 4 neutrons)	3

The atoms of the remainder of the elements are built up in a similar way. The atomic number gives the charge on the nucleus and also the number of planetary electrons; the atomic mass minus the atomic number gives the number of neutrons in the nucleus.

The Bohr atom

The difficulty with Rutherford's model atom is that, according to the laws of classical physics, it cannot exist. The electrons rotating round the nucleus must have an acceleration towards the nucleus and consequently should radiate energy continuously, spiralling towards the nucleus to provide the energy. Moreover, although our solar system would be completely disrupted if a large star came sufficiently close to it, the atoms of a gas are colliding with each other millions of times a second and have a remarkable stability.

In 1913 the Dane, Niels Bohr, was working with Rutherford in Manchester just after the nuclear atom theory had been put forward. He applied to the nuclear atom the quantum theory, originally propounded by Planck to account for the radiation from a black body and subsequently used by Einstein to explain the photoelectric effect. The quantum theory postulates that energy is radiated as small packets of energy, hf called quanta, where h is a universal constant and f is the frequency of the radiation.

Bohr assumed that the planetary electrons in an atom can exist only in a limited number of stable orbits or stationary states having definite amounts of energy but not emitting radiation, and that radiation occurs only when an electron jumps from one stable orbit to another. He assumed

$$hf = E_2 - E_1,$$

where f is the frequency of the energy radiated when an electron jumps from an orbit of energy E_2 to one of energy E_1.

Fig. 13.25

The hydrogen atom consists of a single electron of charge $-e$ revolving round a nucleus of charge $+e$. Suppose the electron has a mass m, that it revolves in a circle, and that when the radius of its orbit is r its velocity is v (Fig. 13.25). The electrostatic attraction between the nucleus and the electron must equal the centrifugal force; thus

$$\frac{e^2}{4\pi r^2 \epsilon_0} = \frac{mv^2}{r}. \tag{1}$$

ALBERT EINSTEIN   NIELS BOHR

ALBERT EINSTEIN (1879–1955) *was born of Jewish parents at Ulm, Germany. His family moved to Milan while he was a boy and he was educated at Zurich. He became a minor official in the patent office at Berne and, in the year 1905, published his Special Theory of Relativity, the theory of the equivalence of mass and energy and the explanation of the photoelectric effect in terms of Planck's quantum theory, all of which have proved of profound and revolutionary significance in modern physics. In 1915 there followed his General Theory of Relativity and from 1919, when British astronomers confirmed his predicted bending of light by the sun, he was the most celebrated scientist in the world. In 1913, at the instigation of Planck, he took a position specially created for him at the Kaiser Wilhelm Institute, Berlin, and, after the first world war, he used to say that the most interesting collection of physicists in the world was to be found in Berlin. With the rise of the Nazis, however, he was forced to resign, if only to spare Planck, the Rector of the Institute, the humiliation of dismissing him because of his non-Aryan origins. In 1933 he was appointed a research professor at the Institute for Advanced Study at Princeton, and in 1940 he became a US citizen. Since two entirely different theories, his own theory of relativity and Newton's theory of gravitation, could account for the same facts, he took the view that a physical theory is a free creation of the imagination, which fits the facts as clothes fit the body.*

NIELS BOHR (1885–1962), *after taking his doctorate at the University of Copenhagen in 1911, spent some time in the Cavendish Laboratory, under J. J. Thompson, and then under Rutherford at Manchester University. In 1913 he applied the quantum theory to Rutherford's nuclear model of the hydrogen atom and thereby opened a new era in atomic physics. In 1920 the Copenhagen Institute for Theoretical Physics was built for him and it became the leading centre of theoretical atomic research in the world. Many pupils flocked to it, among them Heisenberg and Dirac. Bohr attacked the problem of the stability of the nucleus, utilizing with great success the analogy of a charged liquid droplet having surface tension. During the second world war he was active in assisting refugees to escape from Nazi Germany and in 1943, receiving a hint from the underground that he was about to be arrested, escaped with his family to Sweden. He was flown from Sweden to England in the bomb bay of a Mosquito plane, to assist in the making of the atomic bomb. He and his twin brother Harold were footballers of international calibre.*

At this point Bohr made a further assumption which seems purely arbitrary; on the new wave mechanics, however, its meaning becomes more understandable. He assumed that the angular momentum of the electron, mvr, is always an exact multiple of $h/2\pi$; thus

$$mvr = \frac{nh}{2\pi} \tag{2}$$

when n is an integer called the quantum number. The angular momentum is then said to be quantized.

The next step is to find the energy of the electron in its orbit. The potential energy of the electron is the work done in bringing it from infinity to its orbit and this is $-(e^2/4\pi r\epsilon_0)$ (see p. 181); the negative sign indicates that work is done by the electron as it approaches the oppositely charged nucleus. The kinetic energy of the electron is $\frac{1}{2}mv^2$ and from equation (1) this is equal to $\frac{1}{2}(e^2/4\pi r\epsilon_0)$.

$$\therefore \text{ Total energy of electron} = \text{k.e.} + \text{p.e.}$$

$$= \frac{1}{2}\frac{e^2}{4\pi r\epsilon_0} - \frac{e^2}{4\pi r\epsilon_0} = -\frac{1}{2}\frac{e^2}{4\pi r\epsilon_0}.$$

From equation (1)

$$mv^2r = \frac{e^2}{4\pi\epsilon_0}.$$

Squaring equation (2)

$$m^2v^2r^2 = \frac{n^2h^2}{4\pi^2}.$$

Dividing

$$r = \frac{n^2h^2\epsilon_0}{\pi e^2 m},$$

$$\therefore \text{ Energy of electron} = -\frac{1}{2}\frac{e^2}{4\pi r\epsilon_0}$$

$$= -\frac{e^4 m}{8n^2h^2\epsilon_0^2}.$$

$$\therefore hf = E_2 - E_1 = \frac{e^4 m}{8h^2\epsilon_0^2}\left(\frac{1}{n_1^2} - \frac{1}{n_2^2}\right),$$

$$f = \frac{e^4 m}{8h^3\epsilon_0^2}\left(\frac{1}{n_1^2} - \frac{1}{n_2^2}\right).$$

In 1884 Balmer, a Swiss schoolmaster, had discovered a formula representing the series of visible spectral lines of hydrogen. The formula can be written in the form

$$f = R \left(\frac{1}{2^2} - \frac{1}{m^2} \right),$$

where m is an integer greater than 2. The value of R, called the Rydberg constant, was known from the measured frequencies of the spectral lines. Bohr was able to calculate the value of the constant corresponding to R in his formula, i.e. $e^4 m / 8 h^3 \epsilon_0^2$, from the known values of e, m and h. The agreement was perfect within the limits of experimental error.

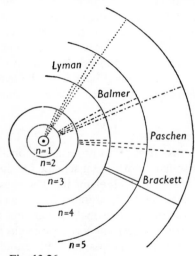

Fig. 13.26

Bohr explained the emission of the visible spectral lines of hydrogen as due to electrons jumping from outer orbits to the second orbit (Fig. 13.26). Other spectral series for hydrogen were also known, the Lyman series in the ultra-violet, the Paschen and the Brackett series in the infra-red. These are due to electrons jumping into the first, third and fourth orbits respectively.

The agreement of Bohr's theory with experiment was so convincing that its revolutionary break with classical physics had to be accepted. Such a break required boldness and genius of the highest order. Rutherford himself said that he regarded Bohr's theory as one of the greatest achievements of the human mind.

Excitation potentials

In the normal condition of the hydrogen atom the electron is in its innermost orbit, for which $n = 1$. As a result of collision, say in a discharge tube, the electron may be knocked into other orbits for which $n = 2$ or 3, etc.: the atom is then said to be excited. The electron will jump back to the innermost orbit, possibly in one jump or in stages, giving out the appropriate radiation. If the electron is completely removed the atom is said to be ionized.

A most satisfying confirmation of Bohr's theory of energy levels may be made by projecting a stream of electrons into rarefied hydrogen. Electrons of small energy collide elastically with the hydrogen atoms and lose no energy, but if the energy of the electrons is increased to $1·6 \times 10^{-18}$ J, which is the difference in the energies of the first and second orbits, then the first spectral line of the Lyman series flashes out. The electrons projected into the gas have just sufficient energy to knock electrons from the first to the second orbits in the atoms and the first Lyman line is emitted during the jump back from the second to the first orbit. Similarly with electrons of greater energies, other lines can be produced.

Electron shells in the atoms

Assuming that the electron structure of an atom is responsible for its chemical properties it is possible to work out a scheme for the arrangement of the electrons in the atoms of all elements. The most inactive elements are the rare gases, helium (atomic number 2), neon (10), argon (18), krypton (36), xenon (54), and it is reasonable to assume that these have a very stable electron structure. Atoms having one electron more than these, lithium (3), sodium (11), potassium (19), rubidium (37), caesium (55), are all very active chemically and have similar properties to each other. The same is true of atoms having one electron less, fluorine (9), chlorine (17), bromine (35), iodine (53).

Atoms are therefore believed to be built up of shells of electrons. When these shells are completely filled the element is inert and when the outermost shell has one electron too many or too few the element is particularly active chemically. Successive shells are filled when they contain $2n^2$ electrons where $n = 1, 2, 3$, etc.: the K shell contains 2 electrons, the L shell 8, the M shell 18, the N shell 32, and the O shell 50. Thus helium has the K shell filled, and neon the K and L shells filled. To explain the case of argon it is necessary to assume that the M shell is divided into sub-shells and a similar assumption is necessary in the case of the N and O shells.

The alkali metals, for example sodium, give very simple line spectra. Sodium, atomic number 11, has complete K and L shells and an additional single electron in the M shell. Bohr assumed that this outermost electron is responsible for the emission of the spectral lines in a manner similar to the hydrogen atom. It is called the valence electron, and is primarily responsible for the chemical properties of sodium.

The origin of X-ray spectra

X-rays are produced when atoms are excited by collision with very fast-moving electrons. Deep lying electrons in the K and L shells are knocked out of their orbits and X-rays are emitted when they jump back. The K series of X-ray lines are caused by electrons jumping back from other shells into the K shell – usually from the L shell, and the L series by a jump into the L shell from an outer shell – usually from the M shell. The regular shift in the lines from element to element, observed by Moseley, is due to the steady increase in charge on the nucleus as we pass from element to element in the table.

Quantum mechanics

After a decade of triumph evidence started to accumulate that something was wrong with the Rutherford–Bohr atom. Small discrepancies between theory and experiment began to appear even for the simplest one-electron atom, hydrogen, while for more complicated atoms, with several optically active electrons in the outer shell, the theory gave totally wrong predictions. Bohr had also been obliged to invent some arbitrary rules, called selection rules, to avoid the prediction of lines not present in the spectra of sodium and the other hydrogen-like atoms. Moreover, his theory rests on the assumption that the angular momentum of the outer electron is quantized but this hypothesis seems quite unjustified and cannot be explained by the theory. Nonetheless, the remarkable successes of Bohr's theory clearly indicated the presence of some new, as yet ungrasped, truths.

For the second time in two decades the way stood clear for minds of genius to create a completely new physical theory. Einstein was led to his theory of Special Relativity (1905) by the surprising fact that however fast an observer recedes from a source of light, the light from that source will always overtake him at the same relative velocity. As Michelson and Morley had shown in 1887 the velocity of light measured by an observer is independent of the observer's motion. Now once again classical ideas dating back to Newton were at an impasse and a new theory, called the Quantum Theory, was about to be developed.

In 1922 de Broglie suggested that just as light can exhibit particle-like properties in the photoelectric effect, so particles might sometimes behave like waves. Light behaving like a particle has an energy given by Planck's relation as we saw in Chapter 10:

$$E = hf.$$

It also has a momentum

$$p = \frac{E}{c}$$

as can be shown from the Special Theory of Relativity.

The momentum can be written in terms of the wavelength, λ, of the light:

$$p = \frac{E}{c} = \frac{hf}{c} = \frac{h}{\lambda}.$$

De Broglie suggested that a particle should similarly have an associated frequency and a wavelength given by the formulae

$$f = \frac{E}{h}, \qquad \lambda = \frac{h}{p} = \frac{h}{mv}.$$

As a consequence electrons accelerated through a p.d. of 100 V should behave like waves of $\lambda = 1 \cdot 22 \times 10^{-20}$ m, about the same as for X-rays.

In 1927 Davisson and Germer diffracted a beam of electrons from the surface of a nickel crystal, in a manner similar to the Bragg experiment for X-rays, and obtained a value for the wavelength agreeing with

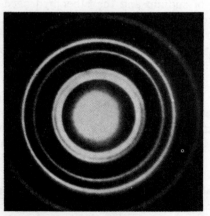

Fig. 13.27. Diffraction of electrons by a thin metal film.

de Broglie's theory. The following year G. P. Thomson, the son of J. J. Thomson, passed electrons through very thin gold or silver foil, an experiment analogous to the von Laue method for X-rays, and obtained diffraction patterns similar to Fig. 13.27.

This startling confirmation of the wavelike properties of particles led, in the same year, to the famous Uncertainty Principle of Heisenberg. It states that it is in principle impossible to be certain both of the position of a particle and its momentum. If one is known with certainty the other will be completely uncertain. This can be understood if we think of a particle as a 'packet of waves'. The particle momentum is given by the wavelength of the particle waves and if we are to measure a wavelength accurately we need a long monochromatic wave-train as in

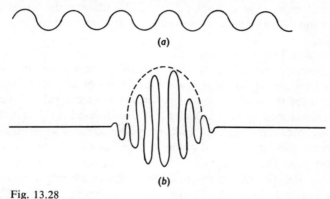

(a)

(b)

Fig. 13.28

Fig. 13.28 (a). However, if we are to be certain of the particles position the waves must be bunched in a tightly confined packet (Fig. 13.28 (b)). The two requirements are contradictory and the particle cannot be at the same time represented both by a long monochromatic train and a tight packet of waves.

The main feature of Quantum Mechanics is the incorporation of this idea of incompatible variables, that is quantities like position and momentum which cannot be simultaneously measured. This is done by replacing the algebraic variables for position and momentum, x and p, of classical theory by two new quantities $\hat{x}$ and $\hat{p}$ called operators. These have the interesting property that $\hat{x}\hat{p}$ does not equal $\hat{p}\hat{x}$, unlike classical theory where of course $xp = px$. The operators stand for the operation of an experimental measurement and because the exact measurement of a particle's position would force it into a tight wave packet, while the exact measurement of momentum would leave a long

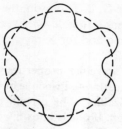

Fig. 13.29

wave-train, so the symbol $\hat{x}\hat{p}$, which stands for the measurement of momentum followed by position resulting in a tight wave-packet, cannot equal the symbol $\hat{p}\hat{x}$ which stands for the measurement of position followed by a measurement of momentum leaving a long wave-train.

Two different mathematical representations, or pictures, of these operators were developed. In 1925 Heisenberg developed his matrix mechanics in which the operators are arrays of number called matrices, which are multiplied together according to special rules. An alternative was developed by Schroedinger in 1926. He used the differential operators of ordinary calculus and derived his famous wave equation for a particle moving in a fixed potential. His equation can be solved for the electron in the hydrogen atom; the solutions are a set of stationary waves and the quantization of the angular momentum is a natural consequence. We can illustrate this by imagining the orbit of the electron round the nucleus to be a stationary wave, Fig. 13.29. Then the length of the path of the electron round the nucleus, $2\pi r$, must be an integral number of wave-lengths $n\lambda$. Since.

$$\lambda = \frac{h}{mv} \quad \text{we have } n\lambda = \frac{nh}{mv} = 2\pi r.$$

This is just Bohr's postulate that the angular momentum, mvr, must be $nh/2\pi$.

We have noted that a particle is represented by a wave with frequency related to the particle energy, and wavelength related to the particle momentum. We might ask what is the meaning of the amplitude of these waves. It was suggested by Born that the square of the amplitude of the electron waves is proportional to the probability of finding the electron in a unit volume at that place. Thus the electron in the hydrogen atom may be pictured as a ball of mist, the probability of the electron's presence at any point being proportional to the density of the mist.

Nuclear reactions

In 1919 Rutherford found that when nitrogen was bombarded with α-particles from radium C, scintillations could be obtained on a fluorescent screen far beyond the range of the α-particles. The particles causing the scintillations were deflected by a magnetic field and shown to be protons, that is, nuclei of hydrogen. Rutherford concluded that, when the nucleus of a nitrogen atom receives a direct hit from an α-particle travelling at high speed, it disintegrates to form an isotope of oxygen. Such processes are known as nuclear reactions and are a useful means of determining the structure of nuclei and the nature of the forces between them. The reaction is represented as follows:

$$\underset{\text{nitrogen}}{^{14}_{7}\text{N}} + \underset{\alpha\text{-particle}}{^{4}_{2}\text{He}} = \underset{\text{oxygen}}{^{17}_{8}\text{O}} + \underset{\text{proton}}{^{1}_{1}\text{H}}$$

Blackett, using a Wilson cloud chamber, was able to photograph the effect (Fig. 13.30).

Rutherford's experiment was the first artificial transmutation of an element – the old dream of the alchemists. He and Chadwick showed that about a dozen of the lighter elements can be transmuted, like nitrogen, by bombardment with α-particles.

Now 1 mg of radium gives off 37 000 000 α-particles per second,

Fig. 13.30. A photograph showing the artificial disintegration of nitrogen. The almost parallel tracks are those of α-particles in nitrogen. One α-particle has entered a nitrogen atom which has ejected a proton (the long track going to the left) and become oxygen (the short track veering to the right).

but, owing to the very large spaces inside the atoms of the substances bombarded, only one α-particle in a million reacts with a nitrogen nucleus. Accordingly, more copious sources of fast-moving particles were devised. In 1932, Cockcroft and Walton, working with Rutherford in the Cavendish Laboratory, Cambridge, accelerated protons in a discharge tube through a p.d. of 400 000 volts. They were able in this way to transmute the light elements lithium and boron. Lithium breaks up into two α-particles of very high speed, indicating a large release of energy derived from a slight annihilation of mass.

$$\text{$_3^7$Li} + \text{$_1^1$H} = 2\text{$_2^4$He}.$$

Particle accelerators

Only light elements can be transmuted by the comparatively small p.d. used by Cockcroft and Walton: the large positive charge on the nucleus of heavier element causes a large repulsion of the bombarding particles and reduces the chance of a direct hit. Very much faster protons, deuterons ($_1^2$H), or α-particles can be produced by a machine called a cyclotron, invented by Professor Lawrence of the University of California.

The principle of the cyclotron is shown in Figs. 13.31 and 13.32. Lawrence is said to have invented it because he found difficulty in following a proof in a German periodical that it was unworkable. Two hollow, metal, half-cylinders D_1 and D_2, called dees, like a pill-box sliced in two, are contained in a larger evacuated vessel and insulated from each other. If a beam of protons is required, a little hydrogen is admitted and some of the gas is ionized by the hot tungsten filament, F. An alternating p.d., of say 100 kV, is set up between the dees. A proton formed near the centre will be accelerated across the gap between the dees from whichever of them is positive at the moment to the negative

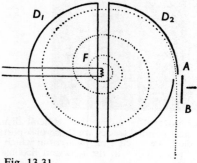

Fig. 13.31

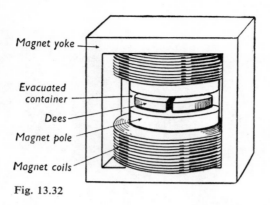

Magnet yoke

Evacuated container

Dees

Magnet pole

Magnet coils

Fig. 13.32

one. A magnetic field at right angles to the paper, set up by a very large and powerful electromagnet, causes the proton, now travelling at a uniform speed inside a dee, to traverse a circular path. If, by the time it is about to cross the gap again, the p.d. between the dees is reversed, it will be accelerated through another 100 kV. Since its velocity is now greater it will travel in a circle of a larger radius, but the times it takes to complete a half-circle is the same as before; its increased speed is exactly compensated by the increased length of its path. Thus the proton can be made to traverse a spiral and if it does 200 turns it acquires a velocity equivalent to that produced by a p.d. of $200 \times 100\,000 = 20$ MV.

On emerging at A the paths of the protons are straightened out into a beam by the attraction of the deflecting plate B.

When the speed of the protons becomes very high, however, their masses increase appreciably, as predicted by the theory of relativity, and this causes them to arrive a little late at the gap between the dees. For a particular frequency of the p.d. between the dees there is only one correct or stable orbit, corresponding to a particular proton mass. If the frequency is reduced, the radius of the stable orbit increases and hence the velocity of the proton can be increased. Thus, by varying the frequency, the energy of the protons can be increased to the maximum, and this is the principle of the synchro-cyclotron (see Fig. 13.33).

In yet another modification of the cyclotron, known as the synchrotron, the particles are accelerated in an orbit of constant radius by increasing the magnetic field. This has the advantage of reducing the size of the magnet required, and hence of reducing the very high cost of these machines. The synchrotron at Batavia Illinois which started operating at 200 G V in 1971 will eventually attain energies of at least

Fig. 13.33. The 4-m synchrocyclotron at Liverpool University has an energy of about 5 MeV. The machine at Batavia has a ring diameter of about 2 km!

500 G eV and if the technology of superconducting magnets allows, energies of 1000 Gev may become possible. A machine of similar size is planned as a European cooperative effort and will be built at the CERN site at Geneva. Several smaller machines, in the 30–70 GeV range, are operating in the USA, Europe and the USSR. Such energies are too large for probing the nucleus but rather examine the structure of individual neutrons and protons. This is the province of physics known as Elementary Particle Physics.

The neutron

From the bombardment of beryllium with α-particles Sir James Chadwick discovered in 1932 a particle with the same mass as a proton but of zero charge, called the neutron:

$$\underset{4}{9}\text{Be} + \underset{2}{4}\text{He} = \underset{6}{12}\text{C} + \text{n}.$$
$$\text{neutron}$$

The radiation from beryllium had been investigated by Bothe and Becker in Germany, and by M. and Mme Curie-Joliot in France and was thought by them to be γ-rays.

Artificial radioactivity

Although they missed narrowly the discovery of the neutron, M. and Mme Curie-Joliot a little later discovered induced radioactivity, for which they were awarded a Nobel prize. They found that aluminium bombarded with α-particles gave rise to a radioactive isotope of phosphorus:

$$\underset{13}{27}\text{Al} + \underset{2}{4}\text{He} = \underset{15}{30}\text{P} + \text{n}.$$

The phosphorus breaks up into silicon and has a half-life period of 2 or 3 minutes.

Several hundred artificially induced, radioactive isotopes are now known. One of the most important is radioactive sodium, $\underset{11}{24}\text{Na}$, which has a half-life period of 15 hours and emits β- and γ-rays as it disintegrates into magnesium. This has therapeutic value, as a substitute for radium, in diseases such as cancer. It can be made easily with a cyclotron by bombardment of ordinary sodium with deuterons:

$$\underset{11}{23}\text{Na} + \underset{1}{2}\text{D} = \underset{11}{24}\text{Na} + \underset{1}{1}\text{H}.$$

Artificially prepared radioactive substances, particularly radioactive phosphorus, have already been used in biological research. The passage of these atoms through the body can be traced by their radioactive effects.

Fission

Before 1939 all known nuclear changes had consisted of the removal of a small mass from the nucleus in the form of an α-particle, a proton or a neutron. In that year Hahn and Strassmann discovered that the nucleus of the uranium isotope, ^{235}U, when bombarded with slow neutrons, splits up into two nearly equal parts. Hahn and Strassmann did not realize what was happening; it was Frisch and Meitner who suggested nuclear fission. Several different nuclei have been identified as the result of the fission, and all that can be said is that the nucleus splits into parts with masses in the approximate ratio 5:7. A comparatively large amount of mass is annihilated in the process and hence a large amount of energy is released. At each fission about three neutrons are emitted. The phenomenon may be represented as in Fig. 13.34.

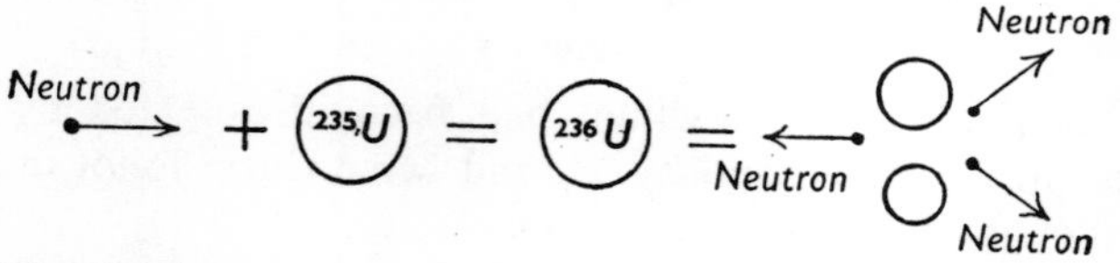

Fig. 13.34

The atomic bomb

It was at once realized that a chain process in a mass of pure ^{235}U is possible; once one nucleus has been split, three others will be split by the neutrons released, then nine others and so on, providing an explosive of unparalleled violence. In November 1941 the American Government was advised by its scientific experts that an atomic bomb could be produced in three or four years.

Natural uranium contains only 0·7 per cent of ^{235}U; the much more abundant isotope is ^{238}U. Two of the main problems connected with the manufacture of the bomb were the production of ^{235}U in quantity and the control of the explosion. The two isotopes of uranium may be separated by two methods: (1) the gaseous diffusion method depending on the slightly more rapid diffusion, through a porous barrier, of lighter isotopes than of heavier; (2) the electromagnetic method, in principle the same as that of the mass spectrograph described on p. 297. Both of these methods when used in the laboratory produce minute quantities of ^{235}U. The idea of their application on an industrial scale would, in peace time, have been regarded as laughable; the production of the atomic bomb cost the USA £500 000 000.

^{235}U does not explode as it is made, despite the fact that there are always stray neutrons present. Neutrons, being uncharged, are undeviated by the charges in the atom, and do not cause fission until they reach a nucleus in their direct path. They must travel a considerable distance, on an average, before being effective. There is therefore a critical size for a bomb, and until this size is reached neutrons escape from the mass of ^{235}U without causing cumulative fission. The explosion may be set off by bringing together two pieces of ^{235}U each too small to explode by itself.

The thermo-nuclear bomb

At a temperature of about 20 million K, such as is found in the sun and the stars, helium can be formed from hydrogen and there is an enormous release of energy due to a destruction of mass. At this temperature the thermal speeds of the hydrogen nuclei are so great that they can overcome each other's electrical repulsion and come into sufficiently close contact to combine. The temperature provided by the uranium bomb is high enough to start this reaction, the bomb acting as a detonator; the energy liberated by the reaction raises the temperature still further and accelerates the reaction. One reaction which has been used is the fusion of lithium and deuterium, made available in the form of lithium deuteride.

Nuclear reactors

Nuclear fission in pure uranium, consisting of a natural mixture of ^{235}U and ^{238}U, can be made to proceed, and its rate controlled, in what is known as a nuclear reactor. Nuclear reactors release energy for a very long period of time and are being designed to act as the source of energy in electric power stations and for propelling submarines and surface ships.

The reason why nuclear fission does not proceed spontaneously in natural uranium is that the neutrons released by the fission of ^{235}U, of which there is always a small amount occurring spontaneously, are absorbed by ^{238}U (which is 140 times more abundant than ^{235}U) and hence are prevented from causing further fission. On absorbing a neutron ^{238}U becomes ^{239}U; from this a β-particle is ejected giving a new element, neptunium, of atomic number 93, and then a further β-particle is ejected giving plutonium of atomic number 94:

$$^{238}_{92}\text{U} + n = {}^{239}_{92}\text{U} = {}^{239}_{93}\text{Np} + e,$$
$$^{239}_{93}\text{Np} = {}^{239}_{94}\text{Pu} + e.$$

If the neutrons released by the fission of ^{235}U are slowed down by elastic collision with suitable nuclei, some of them are not absorbed by ^{238}U and they can therefore cause further fission of ^{235}U. Since there is little change in the speed and kinetic energy of the neutrons at an elastic collision with a heavy nucleus, the light elements are the most effective agents in slowing them down; such elements are called moderators and the most suitable are graphite and heavy water.

The first reactor was built in Chicago in 1942. It consisted of a lattice of lumps of uranium (about 5 tons) and graphite bricks to act as moderators. Cadmium rods, which absorb neutrons readily and hence can be used to control the reaction, passed through ten slots in the reactor. The reactor stopped working when all the rods were pushed home; it was started again by drawing out all the rods except one and then by pulling out the last rod gradually.

During the working of the reactor ^{235}U undergoes fission and is gradually used up. However, plutonium is being formed and plutonium is fissile like ^{235}U, but less plutonium is formed than ^{235}U is used up, so that there is a steady fuel loss. The plutonium produced in a reactor can be separated chemically from the uranium since it is a different element – a much easier process than the separation of isotopes. Thus reactors breed material which can be readily extracted for atomic bombs and the second bomb dropped on Japan was constructed of plutonium.

Besides the above type of reactor, known as a 'slow' reactor, there is another type known as a 'fast reactor'. The latter has a core of nearly pure fissile material surrounded by a 'blanket' of ^{238}U or thorium. Since the core contains no ^{238}U it requires no moderator to slow down the neutrons; some neutrons cause further fission in the core but most are absorbed by the blanket, thereby giving rise to new fissile material— plutonium or, in the case of thorium, ^{233}U, which is also fissile. The fast reactor therefore breeds more fissile material than it consumes but it requires a core of pure fissile material which is at present scarce.

Cosmic rays

At the beginning of this century several physicists were intrigued by the inevitable slow leakage of charge from an electroscope, despite all efforts to make its insulation perfect. The effect was generally ascribed to the ionization of the air by radioactive materials in the earth's crust. However, experiments at the top of the Eiffel tower, and during balloon ascents, showed the effect to increase with increasing height, while electroscopes sunk to a depth of 25 m in lakes showed a decrease in the effect. It became clear that a mysterious radiation was reaching the earth from outer space, and it was given the name of cosmic rays.

The radiation was first thought to be X-rays of very short wavelength but is now known to consist of positively charged particles of immense energy. Charged particles should be deflected by the earth's magnetic field, causing a smaller intensity at the equator than elsewhere, and this latitude effect has been observed experimentally. Moreover the radiation tends to come from a westerly direction, and positively charged particles are deflected from west to east by the earth's magnetic field whereas negatively charged particles would be deflected in the opposite direction.

The two chief ways of investigating cosmic rays are by means of the tracks they make in a Wilson cloud chamber or in a photographic emulsion of extremely fine grain. An energetic charged particle knocks orbital electrons from the atoms in its path and these act as centres of condensation in a cloud chamber or as centres for the growth of grains of silver which, on development, appear under a microscope as black dots in a photographic emulsion.

It has been found that, if a very energetic charged particle makes a direct hit on the nucleus of an atom in its path, an astonishing wealth of new phenomena appear. Cosmic rays have been observed with very high energies (equivalent to those produced by a particle accelerator of 10^{19} V) and together with the proton synchrotrons provide the raw material for Elementary Particle Physicists.

To determine the nature of the primary cosmic rays, so-called before they have become modified by collisions in the earth's atmosphere, it is necessary to send up balloons carrying apparatus to a height of about 25 km; for this purpose, photographic plates are ideal apparatus. At a prearranged height the balloon bursts and the apparatus comes gently to earth by parachute. When the plates are afterwards examined, the mass and speed of the particles can be deduced from the ionization they have caused in the plate, from their range in the plate and from the scattering effects they cause (see Fig. 13.35). It has been found that the primary radiation consists mainly of protons, many with a velocity approaching that of light, and also of the nuclei of heavier atoms such as helium, oxygen, carbon, nitrogen, neon, iron and magnesium. It has been suggested that the percentage of each element in the primary cosmic radiation may correspond to its abundance in the universe.

By the time the radiation reaches sea-level, after passing through the earth's atmosphere, it consists, as a result of collisions, of a greatly increased assortment of particles, known as secondary cosmic radiation.

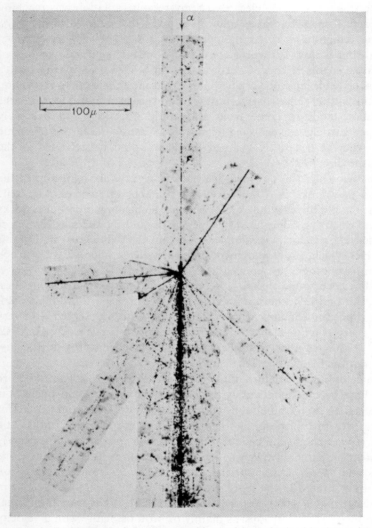

Fig. 13.35. Tracks in a photographic emulsion, obtained in the study of cosmic radiation. An α-particle of energy about 3 TeV (i e. 3×10^{12} eV), coming from the top of the picture, strikes a nucleus, causing it to disintegrate, and a shower of about 140 π-mesons and other particles are created. This is a vivid demonstration of the transformation of kinetic energy into mass.

Several theories of the origin of the primary cosmic radiation have been put forward. One suggests that they are emitted during the atomic transformation in the stars, but this explanation does not account for their huge energies. The energies may be acquired by the passage of the particles through a celestial electric field or the magnetic field of pulsars, the latter acting like a huge cyclotron. (Pulsars are recently discovered celestial objects believed to be rapidly rotating neutron stars – like giant nuclei tens of kilometres in diameter.) However, theoretical difficulties remain and the enormous energies of cosmic rays are still a mystery.

The positron

While experimenting on the passage of cosmic rays through a cloud chamber, across which was a deflecting magnetic field, Carl Anderson of the California Institute of Technology, observed among the secondary rays produced, a particle with the mass of an electron but deflected in the opposite direction, and hence having a positive charge. This positive electron has been termed the positron. Its existence was predicted in 1930 by Dirac but it remained undiscovered until 1932 because it has a very short free life.

When a positron and an electron combine they annihilate each other and produce γ-radiation equivalent in energy to the mass which disappears. The reverse effect, the production of an electron pair from γ-rays, may also be observed when γ-rays of high energy pass through the intense electric field near the nucleus of a heavy atom. Since the energy equivalent of the mass of each electron is 0·5 million electron volts (MeV), the minimum or threshold energy of the γ-ray must be 1 MeV.

Mesons

The nucleus of an atom is believed to consist of protons and neutrons, known as nucleons. The forces inside the nucleus must differ from normal electrical forces; otherwise the mutual repulsion of the protons would cause the nucleus to fly apart and the neutrons would drift away. In 1935 the Japanese physicist Yukawa predicted a new kind of particle, intermediate in mass between the electron and the proton, called the meson, which acts as a kind of mortar to the bricks represented by the nucleons in the nucleus, and which bears the same relation to the field inside the nucleus as do photons to the electromagnetic field.

In 1936 Anderson and Neddermeyer observed a track in their cosmic ray cloud chamber, more strongly deflected in the magnetic field than that of a proton and thicker (i.e. showing more ionization) than that of an electron. This was at the time believed to be the first recorded observation of a meson. It was produced by a direct hit on the nucleus of an atom in the cloud chamber by a high-energy cosmic ray particle.

It later appeared that this was not the particle that was predicted by Yukawa. Professor C. F. Powell and his colleagues at Bristol University, using the photographic emulsion technique, discovered the heavier π-meson (Yukawa's) which decayed into a lighter particle (Anderson's). Mesons can now be produced by bombarding a metal target with protons of high energy from a particle accelerator, which has the great advantage of producing a controlled, intense, well-defined, almost mono-energetic beam, instead of the chance, erratic cosmic ray particles. Anderson's particle is now called a muon and together with the electron belongs to the class of leptons.

Fundamental Particles

Since the Second World War the number of so-called fundamental particles has been growing at an ever-increasing rate and the present number is well into the hundreds with the thousand mark in sight.

The particles are best classified together with the four known types of force or interactions. These are the *strong* interactions responsible for holding the nucleus together and with 'strength' about unity; the *electromagnetic* interaction (the main subject of this text) which binds the electrons to the atom and has strength about 10^{-2}; the *weak* interaction which is responsible for radioactive decay and of strength $\sim 10^{-15}$; and finally the *gravitational* interaction of strength $\sim 10^{-40}$. The photon is the 'mortar' or 'glue' of the electromagnetic interaction and a similar massless particle called the graviton is conjectured as the 'glue' of the gravitational interaction. The electron, positron, two muons and four massless particles called neutrinos make up the class of leptons. They are emitted in radioactive decay processes and seem to be associated with the weak interaction. Another lepton called the W-boson has been conjectured as the 'glue' of the weak interaction but has not yet been observed.

It is in the class of hadrons which are associated with the strong interaction that the greatest proliferation has been seen. The hadrons divide into two subclasses, the mesons and the hyperons. Many more mesons have been discovered since the original triplet of Yukawa mesons the π^+, π^0 and π^-. The π mesons have rest mass energies of about 130 MeV, compared with 0·5 MeV for the electron and 940 MeV for the proton, while next in mass come the K mesons at about 500 MeV and then a great many more ranging up to several GeV.

The word hyperon comes from the Greek and means 'heavy one'. The founder members of the hyperon class are the proton, neutron and their antiparticles. (The positron is the antiparticle of the electron. Antiparticles are characterized by the same mass as the original particle but have the opposite charge and various other labels, known as quantum numbers, change sign.) Several hundred hyperons are known or conjectured again ranging up from the proton mass to several GeV.

Most of these particles are very short lived and exist only for about 10^{-10} to 10^{-20} seconds before decaying into other particles. Only the proton, the electron, the neutrino and the photon are stable against decay.

The situation facing physicists today resembles that facing the early compilers of the periodic table of elements, or those who first tried to understand the complex atomic spectra. It has been suggested that all the known hadrons may be bound states of three new as yet undetected particles, called quarks. Physics would then find itself in a similar situation to that where the elements were to be constructed from electrons and nuclei, or nuclei from protons and neutrons. Alternatively it has been suggested that each particle is a mixture of all other particles. This is called the bootstrap-hypothesis since nature tries to lift herself up by her own bootstraps. Will we find that nature is like a linear chain with every naturally occurring object composed of lesser objects or will we at the last find only a perfect circle?

Summary

1895   Röntgen discovered X-rays.
1896   Becquerel discovered radioactivity.
1897   J. J. Thomson discovered the electron and measured e/m.
1898   Pierre and Marie Curie isolated radium.
1901   Planck's Quantum Hypothesis.
1903   Rutherford and Soddy put forward the Disintegration Theory.
1905   Einstein's special theory of relativity; showed equivalence of mass and energy.
1906   Millikan's accurate determination of the charge of the electron.
1910   Soddy suggested existence of isotopes.
1911   J. J. Thomson's positive-ray experiments.
1911   Rutherford's hypothesis of the nuclear atom.
1912   Von Laue proposed use of crystal for diffracting X-rays.
1912   C. T. R. Wilson's cloud chamber.
1913   Bragg's X-ray spectrograph.
1913   Moseley's experiments.
1913   Bohr applied quantum theory to Rutherford's atom.
1919   Rutherford produced artificial transmutation of the element nitrogen.

1919   Aston's mass spectrograph.
1922   de Broglie's theory of electron waves.
1925   Heisenberg's Matrix Mechanics.
1925   Blackett obtained first photographs, using a cloud chamber, of a nuclear reaction.
1926   Schrödinger's Wave Mechanics.
1927   Davisson and Germer diffracted electron waves.
1927   Heisenberg's Uncertainty Principle.
1928   G. P. Thomson diffracted electron waves.
1930   Dirac's theoretical prediction of positron.
1932   Cockcroft and Walton produced artificial disintegration of lithium by proton bombardment.
1932   Chadwick discovered the neutron.
1932   Lawrence built the first cyclotron.
1932   Urey discovered heavy hydrogen.
1932   Anderson discovered the positron.
1933   M. and Mme Curie-Joliot produced artificially radioactive elements.
1935   Yukawa's theoretical prediction of the meson.
1936   Anderson and Neddermeyer obtained the first photograph of a muon track.
1938   Hahn and Strassmann discovered nuclear fission; explained by Frisch and Meitner.
1942   First reactor.
1945   Atomic bomb.
1947   Bristol Group obtained the first photograph of a π meson track.
1947–1973   Progressively higher energy particle accelerators built and many new particles discovered.

Questions

1. Describe what happens when an electric discharge is passed between electrodes in a tube of gas which is gradually exhausted.

What are the principal properties of cathode rays?   (C)

2. If a stream of electrons each of mass m, charge e, and velocity 3×10^7 m s^{-1} is deflected 2 mm in passing for 10 cm through an electrostatic field of 1800 V m^{-1} perpendicular to their path, find e/m.   (C adapted)

3. Calculate the radius of the circle described by an electron which is projected at right angles to a magnetic field of 0·010 Wb m^{-2} with a velocity of 10^8 ms^{-1}. (Charge and mass of electron are $1·60 \times 10^{-19}$ C and 9×10^{-31} kg respectively.)   (B adapted)

4. Show that the periodic time of a charged particle, moving in a circle in a magnetic field, is independent of its velocity.

Explain how this fact is utilized in the cyclotron.

5. Give the masses, to the nearest whole number, in terms of the mass of the proton, and the charges in terms of the charge of the electron, of the following particles: electron, positron, proton, neutron, deuteron, α-particle.

6. State the rule for the mechanical force on an element of a current-carrying conductor in a magnetic field. Apply the rule to the motion of an electron in a magnetic field. Discuss the nature of the motion when the electron enters the field at right angles.

Describe an experiment whereby the value of the ratio of charge to mass of an electron may be determined.   (N)

7. Find an equation to represent the path of an electron of charge e and mass m projected with velocity v at right angles (i) to a uniform electric field of strength E, (ii) to a uniform magnetic field of flux density B.

A light flexible wire lies in a plane which is perpendicular to a non-uniform magnetic field. If T is the tension and I is the current in the wire, show that it lies along the possible path of an electron in the field, provided $TI = mv/e$.   (CS)

8. A charged particle of mass m and charge e is projected with a velocity v at an angle θ to the direction of (a) a uniform magnetic field, (b) a uniform electric field. Discuss the motion of the particle in each case. How may the velocity of such a particle be measured?   (CS)

9. How does the velocity of an electron depend upon the accelerating voltage to which it is subjected?

Give a general account of three electronic devices employing electrons emitted from a heated cathode and having velocities corresponding to accelerating potentials of tens, hundreds, and many thousands of volts respectively.   (N)

10. Calculate the radius of a water drop which would just remain suspended in the earth's electric field of 300 V m^{-1} when charged with 1 electron. ($e = 1.60 \times 10^{-19}$ coulomb.)   (B adapted)

11. According to Stokes the resistance experienced by a small sphere of radius a, moving through a medium of viscosity η with velocity v, is given by the expression $6\pi a \eta v$. Use this to determine the terminal velocity of an oil drop of density 0.95 g cm^{-3} and radius 10^{-4} cm falling through air of density 0.0013 g cm^{-3}.

The drop carries a charge of $+q$ coulombs and is driven upwards by an electric field of 2000 V m^{-1} with a velocity of 0.036 cm s^{-1}. Calculate the magnitude of q.

Give a brief description of the determination of the electronic charge (e) by experiments with oil drops. (The viscosity of air is 181×10^{-6} g cm^{-1} s^{-1}.)   (N adapted)

12. Describe apparatus by means of which X-rays may be produced. Compare cathode rays and X-rays as regards their nature and properties.   (D)

13. Draw a labelled diagram of some form of X-ray tube and of its electrical connections when in actual use.

Electrons starting from rest and passing through a potential difference of 1000 volts are found to acquire a velocity of 1.88×10^9 cm s^{-1}. Calculate the ratio of the charge to the mass of the electron in coulombs per g.   (N)

14. What is the shortest wavelength of X-rays produced by million-volt electrons? ($e = 1.60 \times 10^{-19}$ C; $h = 6.55 \times 10^{-34}$ J s.)

15. Discuss the properties of the three main types of radiation produced during radiactive changes.   (B)

16. Compare and contrast the conduction of electricity through (a) metals, (b) electrolytes, (c) ionized gases.   (CS)

17. Write an essay on *one* of the following:
   (a) The disintegration of the atom.
   (b) The release of atomic energy.
   (c) The structure of the atom.

18. What are *positive rays*? How can a beam of positive rays be obtained in a discharge tube?

Without deriving any formulae, give a short descriptive account of a method of separating isotopes existing in a beam of positive rays.

The mass of the atom of the neon isotope ^{20}Ne is $3\cdot2 \times 10^{-26}$ kg, and the charge on a singly ionized atom is $1\cdot6 \times 10^{-19}$ C. What is the flux density of a magnetic field in which singly charged ions of ^{20}Ne travelling at 10^6 m s^{-1} describe an arc of radius of curvature $0\cdot36$ m?

Find also the radii of curvature of the arcs described in this field by (a) singly charged ^{22}Ne ions and (b) doubly charged ^{20}Ne ions when the velocity is 10^6 m s^{-1} in each case.   (O 1971)

19. Draw a clear diagram showing the chief features of a cathode-ray oscilloscope tube.

Explain the action of the focusing and brilliance controls, and show how the appropriate voltages are applied.

Describe the use of the cathode-ray oscilloscope (a) as a voltmeter, (b) as a means of demonstrating the wave-form of a steady musical note of fixed frequency and (c) as a method of measuring short time intervals.   (O 1971)

20. Answer *two* of the following:
   (i) Describe a cathode ray tube and explain the principles of its operation. How would you use it to display the current vs. voltage characteristic of a diode rectifier?
   (ii) Outline briefly the experiments you would carry out on the photoelectric effect in order to obtain a value for Planck's constant h. What other physical constants would you need in order to calculate h from your data?
   (iii) Describe how you would attempt to determine the half-life of a radioactive material for which the half-life is known to be about ten minutes. What is the nature of the radiation emitted by this material if most of it can penetrate 1 mm of aluminium, but it is completely absorbed by 1 cm of aluminium?   (O & C 1970)

21. Describe the nature of α, β, and γ radiations.

What is meant by the statement that the stable isotope of gold has an atomic number of 79 and a mass number of 197? A sample of pure gold is irradiated with neutrons to produce a small proportion of the radioactive isotope of gold of mass number 198. What experiments would you perform to examine the radiation emitted by the sample to establish whether it was α, β, or γ radiation? If chemical analysis of the sample subsequently showed that it contained a trace of mercury (atomic number 80) what would you conclude from this about the nature of the radiation from the radioactive gold? What would you expect the mass number of the isotope of mercury present in the gold to be?   (O & C 1968)

22. Give an account of J. J. Thomson's determination of the specific charge e/m of the electron.

Give the theory of the method, and explain how the velocity of the electrons in the beam was estimated.

In an experiment of this type, electrons which are travelling with speed 2×10^7 m s^{-1} at right angles to a uniform magnetic field, produced by a steady current of 5 A circulating through the field coils, are observed to describe paths whose radius of curvature in the field is 0·2 m. Calculate the flux density of the magnetic field.

When an electric field of 1250 V m^{-1} is applied in the appropriate direction, it is found that the above magnetic field is too large to enable an undeviated electron beam to be obtained in the two combined fields. To what value must the current in the field coils be reduced in order to obtain an undeviated electron beam?

(Take the value of e/m to be $1·77 \times 10^{11}$ C kg^{-1}.)   (O 1971)

23. Explain the physical principles involved in Millikan's method for determining the value of the electronic charge e. How did he establish that this is indeed the fundamental 'atom' of electric charge?

A uniform electric field of strength 5×10^5 V m^{-1} acts vertically downwards. A small oil drop carrying a charge $8e$ and situated in the field falls with a uniform speed of 2×10^{-4} m s^{-1}; when the charge on the drop is increased to $11e$, it moves upwards in the field with a uniform speed of 10^{-4} m s^{-1}. Find

(a) the mass of the drop;

(b) the charge it carries when it just remains at rest in the field of 5×10^5 V m^{-1};

(c) the strength of the field in which it would just remain at rest when carrying a charge of $8e$.

(Take the value of e to be $-1·6 \times 10^{-19}$ C.)   (O 1971)

24. The properties of electromagnetic radiation can be explained by regarding the radiation either as a wave motion or as a stream of particles. Which of these two interpretations is better fitted to account for (a) the diffraction of X-rays by crystals, (b) the generation of the continuous X-ray spectrum emitted from the target of an X-ray tube, (c) the emission of visible line spectra, (d) the photoelectric effect, (e) the refraction of light at a plane interface? Discuss *two* of these effects in detail.   (O 1968)

25. Describe an apparatus for determining the ratio of the charge e to the mass m of the electron. Explain how measurements are carried out with the apparatus, and derive the relationship between e/m and the experimentally measured quantities.

Indicate briefly how you would attempt to test whether the particles emitted in the photoelectric effect and in thermionic emission are the same.

When low-energy electrons are moving at right angles to a uniform magnetic field of flux density 10^{-3} Wb m^{-2} they describe circular orbits $2·82 \times 10^7$ times per second. Deduce a value for e/m.   (O & C 1968)

26. State the nature of *alpha, beta* and *gamma* radiations.

Suggest a method by which you could ascertain whether a beam of particles from a radioactive source consists of α-particles alone, β-particles alone, or a mixture of α-particles and β-particles. State what apparatus you would require.

The radioactive isotope of polonium, ^{218}Po, decays with a half-life of 3 minutes.

It emits α-particles and produces a radioactive isotope of lead, ^{214}Pb, which emits β-particles and has a half-life of 27 minutes. On the same axes draw sketch graphs to show how, starting with pure ^{218}Po, (a) the α-particle emission-rate, and (b) the amount of ^{214}Pb present, would vary with time.   (O & C 1971)

27. Outline the evidence for the existence of electronic energy levels within the atom.

Figure 13.36 represents a typical energy-level diagram for the hydrogen atom, with some details omitted. Energies are given in electron-volts, with the zero of energy corresponding to the complete removal of the electron from the atom; wavelengths of the emitted spectral lines are given in nanometres (nm).

(a) Explain why the arrows shown indicate possible emission of radiation.

(b) Why does the spectrum of the radiation emitted take the form of sharp lines of definite wavelength?

(c) Find the energy associated with the level $n = 4$, and the wavelength of the line indicated by the transition P.

(d) Show, without any detailed calculations, that all lines resulting from transitions that end on the level $n = 1$ must be in the ultra-violet.

(O 1969)

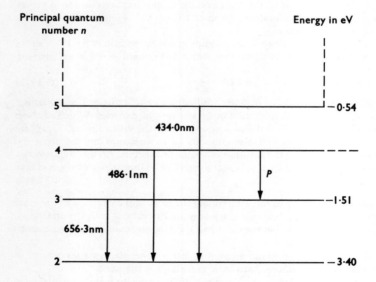

Fig. 13.36

Answers to questions

Abbreviations: watt W, ampere A, coulomb C, volt V, ohm Ω, weber Wb, henry H, farad F, etc.

CHAPTER 1 (*page* 15)

2. $2 \cdot 65 \times 10^7$ m s^{-1}

4. 13 Ω

5. $0 \cdot 44$ Ω

6. $\frac{2}{3}$, $\frac{3}{4}$, 1, $1\frac{1}{3}$, 2, 3, 4, 5, 6 Ω

7. 10 Ω, $1 \cdot 6$ V

8. 1 V, 2 V

9. 3

10. 4 groups, each of 3 cells in series, in parallel

12. $2\frac{17}{32}$ W, $2\frac{1}{4}$ V

13. $\frac{5}{6}$

14. $1 \cdot 59$ kg

15. Yes: cost is $4\frac{1}{2}$p

17. 4180 J kg^{-1} K^{-1}

18. $\frac{2}{3}$, $\frac{1}{3}$, $\frac{1}{3}$, A

19. $1\frac{7}{13}$ V

20. $1 \cdot 31$ V

21. $\frac{21}{82}$ A in AB and BC, $\frac{17}{82}$ A in AD, $\frac{29}{82}$ A in DC

22. $E/8$ A in AB, CB, DC and DA; $E/4$ A in BD; zero in AC

23. $31 \cdot 9$ °C

24. $1 \cdot 07$ mm

25. $\dfrac{I^2 \rho}{2\pi^2 r^3}$, $1 \cdot 129 : 1$

26. 31

27. 58

28. $22 \cdot 7$, $4 \cdot 37$ kW

CHAPTER 2 (*page* 43)

1. $0 \cdot 020$ Ω shunt

2. (*a*) $0 \cdot 0020$ Ω shunt, (*b*) 9980 Ω in series

3. (*a*) 9988 Ω in series, (*b*) $0 \cdot 0012$ Ω shunt

4. 980 Ω in series

5. 10^{-9} A per scale division

8. $0 \cdot 5$ Ω

9. 75 cm, 90 cm, $39 \cdot 5$ cm

12. $0 \cdot 0675$ V

13. $\frac{2}{145}$ A

15. $32 \cdot 05$ km from B

18. $0 \cdot 022$ Ω

19. $2 : 3$

20. $0 \cdot 051$ Ω

21. $8 \cdot 2 \times 10^{-7}$ ohm-m

22. (*a*) $0 \cdot 939$ Ω, (*b*) $1 \cdot 092$ Ω

23. 2500 °C

25. $0 \cdot 20$ Ω

26. $46 \cdot 4$ V

27. $\frac{5}{2}$

29. 5 mV, $5 \cdot 21$ mA, 383 Ω

30. (*a*) $1 \cdot 03$ V, (*b*) $2 \cdot 33$ mA from D to B

Chapter 3 (*page* 70)

2. 1990

3. $17\cdot6 \times 10^{-6}$ Wb m^{-2}

6. $6\cdot76$ mWb m^{-2}, $7\cdot54$ mWb m^{-2}

7. (*a*) zero, (*b*) $4\cdot02\,\mu$Wb m^{-2}; 5×10^{-5} N m^{-1}

8. $0\cdot6'$

9. $4\cdot5$ A

10. 16 N m^{-1}

14. $5\cdot03\,\mu$N, $25\cdot1$ nN m

15. $278\,\mu$N m

17. $\dfrac{0\cdot715\mu NI}{a}$, $< 0\cdot01\%$

18. $88\,\mu$N m

21. (*a*) $0\cdot33$ mN, (*b*) $0\cdot29$ mN $\rightarrow$, (*c*) $0\cdot39$ mN $\leftarrow$

Chapter 4 (*page* 105)

3. $2\cdot5$ V

4. 11 mC

5. $0\cdot25$ Wb m^{-2}

6. *A* will be more sensitive: deflections of *A* and *B* for a given change of flux are 101:15

7. $\frac{1}{3}$ V

8. $1\cdot6$ mV

9. 39 mV

10. $7\cdot5\,\mu$V

11. $1\cdot8 \times 10^{-5}$ Wb m^{-2}

12. 600 r.p.m.

13. (*a*) $0\cdot1$ N, (*b*) zero, (*c*) $0\cdot05$ *N*

14. $4\cdot6 \times 10^{-2}$ N at 23° with horizontal

15. $0\cdot64$ N, $0\cdot256$ N, 25 m s^{-1}

16. $9\cdot4$ mm

17. (*a*) $2\cdot7 \times 10^{-4}$ N m, (*b*) $2\cdot3 \times 10^{-4}$ N m

18. $1\cdot33$ nA

25. $11\cdot5{:}1$, 87 mA, 20 W

27. $4\cdot57$ MW, $22\cdot85$ kA

31. $99\cdot62$ V, $91\cdot25$ V, $91\cdot25\%$

32. (*a*) $152\cdot5$ V, (*b*) $14\cdot4875$ kW

34. (*a*) $37\cdot7$ V, (*b*) $32\cdot6$ V, (*c*) $18\cdot9$ V, (*d*) zero

35. $2\cdot5$ V

36. $19\cdot7$ V, $0\cdot314$ N m

37. $(2\pi^2 r^2 n N\omega I)/R \times 10^{-7}$ V, $\omega/2\pi$ Hz

38. $22\cdot5$ kV

39. $0\cdot5$ H

41. 79 mH

44. $0\cdot59$ div. per μC

45. (*a*) $4\pi \times 10^{-6}$ Wb, (*b*) $8\pi \times 10^{-5}$ V

46. $1\cdot6$ nC div^{-1}, 48 nC

47. 4 Ω, 15 mH, $22\cdot9$ V

49. (*a*) $16\cdot97$ mH, (*b*) $5\cdot66$ mH, (*c*) $0\cdot71$ mC, (*d*) $1\cdot64$ A

Chapter 5 (*page* 128)

1. $0\cdot198$ Wb m^{-2}, $1\cdot33 \times 10^{-4}$ H m^{-1}, 106

3. $0\cdot34$ mC

4. $4\cdot97$ mC

CHAPTER 6 (*page* 144)

1. $2 \cdot 8 \times 10^{-5}$ newton-metre, $1 \cdot 86$ mN
2. 10^{-5} Wb
3. $8 \cdot 1$ mm per 10^{-2} μA
4. $0 \cdot 11$ A

6. $1 \cdot 89 \times 10^{-5}$ Wb m^{-2}
7. 66° $30'$
8. $1 \cdot 91 \times 10^{-5}$ Wb m^{-2}
9. $4 \cdot 93 \times 10^{-5}$ Wb m^{-2}

CHAPTER 7 (*page* 158)

1. 2717 s
2. 11×10^{-5} m^3
3. $1 \cdot 08 \times 10^6$ C
4. $2 \cdot 75$ A
6. 10%
7. $6 \cdot 06 \times 10^{23}$ (or $\times 10^{26}$ if you mean a

kilomole, which apparently most people do not)
$1 \cdot 65 \times 10^{-27}$ kg
11. $2 \cdot 5$ Ω
12. $0 \cdot 95$ A, $3 \cdot 7$ A
14. 5, $9 \cdot 5$p, 94%
15. $\frac{22}{35}$

CHAPTER 8 (*page* 195)

12. $8 \cdot 89 \times 10^{-12}$ F m^{-1}
13. 380 cm^2, $0 \cdot 80$ μC
14. $23 \cdot 6$ div. per μC
15. (*a*) 7 μF, (*b*) $\frac{4}{7}$ μF
   In parallel: charges 100 μC, 200 μC, 400 μC; p.d.'s all 100 V
   In series: charges all $57 \cdot 1$ μC; p.d.'s $57 \cdot 1$ V, $28 \cdot 6$ V, $14 \cdot 3$ V
16. $2 : 1$
17. 60 kV
18. (*a*) $1 \cdot 44 \times 10^{-2}$ J, (*b*) $6 \cdot 8 \times 10^{-3}$ J
19. 200 V m^{-1}
21. $16\ 700$ V m^{-1}, 2×10^{-9} s
22. $4 \cdot 89 \times 10^{-15}$ kg
23. $4 \cdot 2 \times 10^{-4}$ μC
25. 5, $0 \cdot 2$
26. 950 V, $1 \cdot 43 \times 10^{-5}$ J; 3390 V, $5 \cdot 09 \times 10^{-5}$ J
27. 66 μm
28. 2000 J
29. $1 \cdot 11 \times 10^{-5}$ μF

30. $1 \cdot 14 \times 10^{-7}$ F km^{-1}
31. 710 μF
32. $\frac{3}{4}$ μC, $\frac{1}{4}$ μC; $16 \cdot 6$ μC m^{-2}, $49 \cdot 7$ μC m^{-2}
33. $5 \cdot 56 \times 10^{-5}$ J
34. 33 nC
35. (*a*) $1 \cdot 25 \times 10^{-6}$ J, (*b*) $1 \cdot 25 \times 10^{-5}$ J
36. $E_1 = \dfrac{Q\sqrt{3}}{4\pi\epsilon_0 z^2}$, $V_1 = \dfrac{2Q}{4\pi\epsilon_0 z}$,

   $E_2 = \dfrac{Q}{4\pi\epsilon_0 z^2}$ $V_2 = 0$
37. 1000 V m^{-1}, 2000 V m^{-1}, 3000 V m^{-1}
   1000 ϵ, 1000 ϵ
   40 V
38. (*a*) 6000 V, (*b*) $12\ 000\epsilon_0 C$, (*c*) $12\ 000$ V m^{-1}
39. $0 \cdot 565$ N, 1695 V, $0 \cdot 565$ N, $16\ 950$ V
40. (*a*) $8 \cdot 85 \times 10^{-9}$ C, (*b*) $8 \cdot 85 \times 10^{-7}$ J, (*c*) $8 \cdot 85 \times 10^{-2}$ N
41. (*a*) $0 \cdot 736$ K s^{-1}, (*b*) $288 \cdot 736$ K

CHAPTER 9 (*page* 216)

1. 50·1 Hz
2. 2·8 A
3. 283 V
7. 0·502 A, 283 V
9. 0·127 H, 0·53 kW
12. (*a*) 100·8 Ω, (*b*) 1260 Ω
13. (*a*) 3·18 A, 90° lag; (*b*) 5·02 A, 90° lead; (*c*) 3·03 A, 72° 21′ lag, (*d*) 4·54 A, 63° 0′ lead; (*e*) 6·47 A, 57° 6′ lag.
14. 503 kHz

15. 0·16
16. 8·39 V, 16·8 W, 0·95
17. (*a*) 2·875 A, (*b*) 452 V on coil, 229 V on capacitor
20. (*a*) 142 V, (*b*) 100 V, (*c*) 142 V
21. 4·7 A, 89 W, 14·8 V across L, 18·8 V across R
23. (*a*) 3·0 Ω, (*b*) 5·0 Ω, (*c*) 3 A, (*d*) 12 V 9 V, (*e*) 36 W, (*f*) 37°

CHAPTER 10 (*page* 247)

4. 500 °C
7. 2 groups in parallel, each of 3 thermocouples in series, 100 divisions
16. 200 W
18. 3·5 cm, 595 μ Wb m^{-2}

19. $2·5 \times 10^{17}$ s^{-1}, 0·4 μA
21. $6·66 \times 10^{-34}$ Js, 3·34 eV
22. 1·78 mm
23. (*a*) 28·6, (*b*) 4·7 mA/V, (*c*) 5·6 kΩ, all measured near $V_G = -1$, $I_A = 6$ mA

CHAPTER 11 (*page* 273)

6. 150 μA, 270 mV

CHAPTER 13 (*page* 334)

2. 2×10^{11} C kg^{-1}
3. 5·7 cm
10. $1·05 \times 10^{-5}$ cm
11. 0·0114 cm s^{-1}, $8·09 \times 10^{-19}$ C
13. $1·77 \times 10^{8}$ C g^{-1}
14. $1·23 \times 10^{-12}$ m
18. 0·556 T, (*a*) 0·396 m, (*b*) 0·18 m

20. (iii) β
21. β, 197
22. 0·566 mWb m^{-2}, 0·55 A
23. (*a*) $8·16 \times 10^{-14}$ kg, (*b*) 10e, (*c*) 6·25 $\times 10^{5}$ V m^{-1}
25. $1·77 \times 10^{11}$ C kg^{-1}
27. (*c*) 0·87 eV, 1940 nm

Bibliography

W. F. Archenhold, *Electromagnetism and Electrostatics using SI Units* (Oliver and Boyd, 1969).

James J. Brophy, *Semiconductor Devices* (George Allen and Unwin, 1966).

R. G. Hibberd, *Transistor Pocket Book* (Newnes, 1965).

J. Jenkins and W. H. Jarvis, *Basic Principles of Electronics*, Volume 1: *Thermionics*, Volume 2: *Semiconductors* (Pergamon, 1971).

Index